The Future of Sapiens and Verticality

Comparative Anatomy and Posture of Animal and Human Set

coordinated by
Djillali Hadjouis

Volume 6

The Future of Sapiens and Verticality

From Embryo to Life in Space, A New Glance at the Origins and the Future of our Verticality

Edited by

Anne Dambricourt Malassé
Djillali Hadjouis
Sandra Joffroy
Bernard Autet

ISTE

WILEY

First published 2026 in Great Britain and the United States by ISTE Ltd and John Wiley & Sons, Inc.

ISTE Ltd
27-37 St George's Road
London SW19 4EU
UK

www.iste.co.uk

John Wiley & Sons, Inc.
111 River Street
Hoboken, NJ 07030
USA

www.wiley.com

Library of Congress Control Number: 2026934929

British Library Cataloguing-in-Publication Data
A CIP record for this book is available from the British Library
ISBN 978-1-83669-077-1

The manufacturer's authorized representative according to the EU General Product Safety Regulation is Wiley-VCH GmbH, Boschstr. 12, 69469 Weinheim, Germany, e-mail: Product_Safety@wiley.com.

Contents

Preface

P.1. *Sapiens* facing the "wall of verticality"

This collective work, *The Future of* Sapiens *and Verticality*, presents, among other research, the work presented at the first symposium of the FREHOPS[1] (Federation for Research into Human Evolution, Osteopathy and Posture to Serve Health) association, which was held on October 20, 2023, in Paris, at the Institut de Paléontologie Humaine (Institute of Human Paleontology), a foundation established by Prince Albert I of Monaco.

FREHOPS was created in 2018 in order to pool knowledge from different disciplines concerned with the axial verticality of our species, known as "anatomically modern *Homo sapiens*", and by extension, with the verticality of our central nervous system and that of the supporting connective tissues of the bones, fascia and the cardiovascular system. This knowledge focuses on the means by which it is acquired in the primate lineage, as well as on modern-day developmental disorders linked to internal verticality, which begins at the center of the internal base of the skull. This federation sets itself apart through its insistence that this verticality is not locomotive in origin. It is established during human embryogenesis along the rostrocaudal axis of the dorsal chord. This straightening is the result of a very long evolutionary process that resulted in an undeniable internal geometric, craniosacral trajectory, accompanied, at a minimum, by increasing complexities of the brain, particularly the neocerebellum and its connections with the cerebral neocortex. This trajectory, documented by numerous primate fossils and their modern-day descendants, reveals a stability in the modalities of its past evolution, which cannot

Preface written by Anne DAMBRICOURT MALASSÉ, Bernard AUTET, Sandra JOFFROY and Djillali HADJOUIS.

1 See: www.frehops.org.

be explained without the reproducibility of their own principles, the effects of which are, at a minimum, the increasing complexity of neural embryogenesis. Yet, these modalities, whether genetic or epigenetic, continue to be misunderstood, even though they constitute our phylogenetic heritage, that of processes capable of reproducing themselves.

This reproducibility therefore challenges us, here and now, and encourages us to know how to position ourselves on this irreversible curve and to reflect on this inherited destiny, which invites the pooling of knowledge and questions in the face of the geometric limit imposed by the internal verticality that characterizes us as *Sapiens*. Indeed, while this marks the physical limit, and therefore the end of a phylogenetic trajectory of supporting tissues, verticality is not the end of the process. It is, in short, a dynamic epiphenomenon, the result of irreversible processes whose mechanisms, transmitted by gametes at the time of fertilization, still elude scientists. At most, we can infer catalytic properties of self-amplification, or sensitivity to initial conditions (SIC), to account for those of the preceding geometric, architectonic, heterochronic and neurological effects, and their epigenetic consequences. In view of the last 10,000 years of brain, dental and occlusal–postural evolution, these processes have not ceased and we are the result. Neural crest cells and the ways in which their development is regulated by HOX genes provide new insights here.

This curve shows that the straightening of the embryo began with a small inflection approximately 40 million years ago (Ma), from a "prosimian", quadruped and arboreal primate strain, which was then amplified, but in successive steps millions of years apart, up to the first threshold of verticality. This threshold has seen the disappearance of quadrupedalism, while the occasional bipedalism that was already very much present has established itself as the mode of movement of a verticalized central nervous system with the internal tensions of its membrane envelopes, and no longer semi-upright. From this threshold, documented from 4 million years ago by the first hominids sensu stricto[2], the phylogenetic time curve has followed an acceleration up to our so-called "modern" anatomy.

This acceleration, observed but still unexplained, positions us here and now, at the physical limit of internal verticality, and consequently, facing a genuine question for the future of our species: since, without knowing the internal mechanisms of complexification, at least that of the embryonic central nervous system, and to the extent that the process seems to follow an exponential curve, it seems necessary to

2 The introduction of embryogenesis and its rostrocaudal chronology categorically distinguishes the great apes, formerly known as pongids, from the hominids, which group together species with a straightened dorsal chord (*Australopithecus*, *Paranthropus*, *Kenyanthropus*, *Homo* and *Sapiens*).

place the developmental disorders linked to this evolution of embryogenesis, within this perspective of a physical limit to verticalization.

The question of the future of *Sapiens* is therefore that of verticality, and all of its corollaries in psychomotricity, neurophysiology, anatomy, function, posture and energy. It is thus necessary to reposition the process of verticalization of our modern-day species on the phylogenetic curve, in order to anticipate its future effects and to question the relevance of the conquest of all the spaces for which *Sapiens* has become the most "specialized" species, thanks to its technological intelligence and inventiveness, to question, consequently, the role and place of artificial intelligence (AI) in the cultural effects of the process of humanization and its future.

The acceleration of this process has been observed since the first threshold of internal verticality, i.e. for 4 million years. Several lineages have been reproduced, but all have become extinct with the exception of the *Sapiens*: *Australopithecus*, *Kenyanthropus*, *Paranthropus* and all fossil species of the genus *Homo*, with one case that can shed light on these extinction factors, Neanderthal Man, by seeking to understand the process that made it a less verticalized species than the oldest species of the genus *Homo* from which it is derived. Extinction is also one of the consequences of the processes of evolution. The Anthropocene, with global warming and the mass extinction of plant and animal species, presents the daily questioning of this, a global phenomenon of a speed on the scale of geological times, which humanity has never before faced.

The work is divided into two parts. The first part examines "paleontological perspectives" and the second part examines "current perspectives: from the fetus to life in space".

Palaeontology shows, with the example of the Neanderthal, how much verticality presents itself as an evolutionary problem probably in terms of energy balance (Chapter 1: Anne Dambricourt Malassé, Djillali Hadjouis, Sandra Joffroy and Bernard Autet), with consequences for their cardiovascular system (Chapter 2: Sandra Joffroy, Anne Dambricourt Malassé and Djillali Hadjouis). The problem is enriched by an attempt at the virtual reconstruction of one of the oldest *Homo erectus* skull bases in North China – a 780 ka *Sinanthropus* – by Tan-Nhu Nguyen (Chapter 3: Tan-Nhu Nguyen, Anne Dambricourt Malassé, Marie-Christine Ho Ba Tho, Fabienne Lallouet and Tien-Tuan Dao). Lastly, the bone and its characterization (analysis of its mechanical, physical, chemical and physico-chemical properties) are honored with Marie-Christine Ho Ba Tho (Chapter 4).

The second part begins with the hidden side of life, "forgotten" by paleoanthropology, which fills the great absence: what humanizes organic hominization. Humanization begins with intrauterine development through the relationship between the fetus, its environment and its mother; postnatal posture depends on it, as Catherine Dolto shows with clinical observations of haptonomy: "We use the word 'womb' in haptonomy to differentiate it from the uterus as a hollow muscle, empty or refusing the child, and the inseparable whole formed by the contents of the pelvis [...] the child, installed in a house of muscles, fascia and ligaments of which it perceives the most minimal modifications" (Chapter 5). We can add to this knowledge, which is so fundamental and yet still ignored by classical paleoanthropology, the end of the "obstetrical dilemma", with July Bouhallier, who develops and calls for awareness of a new paradigm to put an end to preconceived and erroneous social images, regarding the bodies of pregnant and parturient women, which are detrimental to them (Chapter 6).

How can a child keep or learn verticality when the neurological capital essential to postural motor control is defective? The question is posed by two rare clinical cases in the practice of neuromotor rehabilitation, presented by Véronique Leroy-Malherbe, Élise Pineau and Virginie Pineau, with their conclusion changing the way this disability is viewed: "It is not wrong to compare our two children to extraordinary individuals, those for whom walking has become an art, those who call upon this same ambition of bodily demands in the face of their massive deficit" (Chapter 7).

Postural evolution is also particularly well illustrated by dental occlusion. These links are at the origin of the concept of orthoposturodontics by Michel Clauzade, whose clinical studies focus on a statistical base of 853 patients. The relationship between intercuspidation and postural control is a finding too often ignored by schools of dento-maxillofacial orthopedics, while statistics point toward the establishment of a new postural balance in skeletal class II (recessed mandible), with the trigeminal nerve presenting the common thread of unity between occlusal balance, nervous system and posture (Chapter 8: Michel Clauzade and Numa Clauzade).

Lastly, daily practice in manual therapy and rehabilitation applied to astronaut posturology contributes to space medicine research, confronted with a phenomenon known as "deconditioning" under the effect of weightlessness, with physical, physiological and psychological consequences, similar to accelerated aging. In return, space medicine provides some clues to note that in fact, internal verticality is the axis of individual humanization, and that it remains a phenomenon that is inseparable from the Earth's gravitational force, raising the question of the viability of human beings sent on space missions over long distances and durations such as to Mars (Chapter 9: Denis Ducommun and Loïc Treffel).

P.2. Disciplines called upon to converge

The choice of the title of this collective work is the result of the meeting, more than 20 years ago, between two disciplines that still have few opportunities to exchange their knowledge: human paleontology and osteopathy.

"The future of verticality" expresses, very succinctly, a new awareness since Jean-Baptiste de Lamarck (1802), at the very beginning of the 19th century, made an unprecedented breakthrough in the history of humanity by understanding the anatomy of our species as the consequence of a series of transformations of extinct animal species forming the order of primates.

Since then, fossils of skeletons and teeth, as well as biology and biomechanics knowledge, have corroborated Lamarck, emphasizing his functional interpretation of transformations, essentially the need to survive environmental changes impacting locomotion. The problem emerged of itself as being that of the locomotor (or appendicular) system, with quadrupedalism and arboricolism having disappeared due to the use of bipedalism, imposed by the expansion of open environments such as the Savannah, thus becoming exclusive. Concerning ourselves with the future of our species would thus involve worrying about bipedalism having a future in itself, whereas the contemporary evolutionary paradigm has never conceived it as resulting from phylogenetic determinism.

The situation is different when the study takes into account the dynamic trajectories of development from the embryonic phase in its amniotic environment: human bipedalism then appears as one of the many postnatal effects of an evolution involving the entire organism built around the embryo's rostrocaudal axis and manifesting itself in a reproducible manner according to mechanisms of increasing complexification. This phylogenetic determinism appears to be distinguished by memorization properties and could reproduce from our evolutionary stage, according to *stimuli* whose parameters we do not yet know. Anticipation then takes on meaning and may find certain relevance in bringing together disciplines seeking answers to emerging questions or those that offer new therapies to be shared between disciplines unaccustomed to meeting.

FREHOPS thus sets itself apart from associations dedicated to the study of the evolution of hominids, by recognizing the evolutionary problem that unites clinicians and paleoanthropologists. This problem is the acquisition of internal verticality, the culmination of a phylogenesis marked by a geometric orientation, that of the straightening of the craniocaudal axis of the embryo, more precisely of the dorsal chord, in close connection with the complexification of the nervous system and the tensions that animate its dynamics.

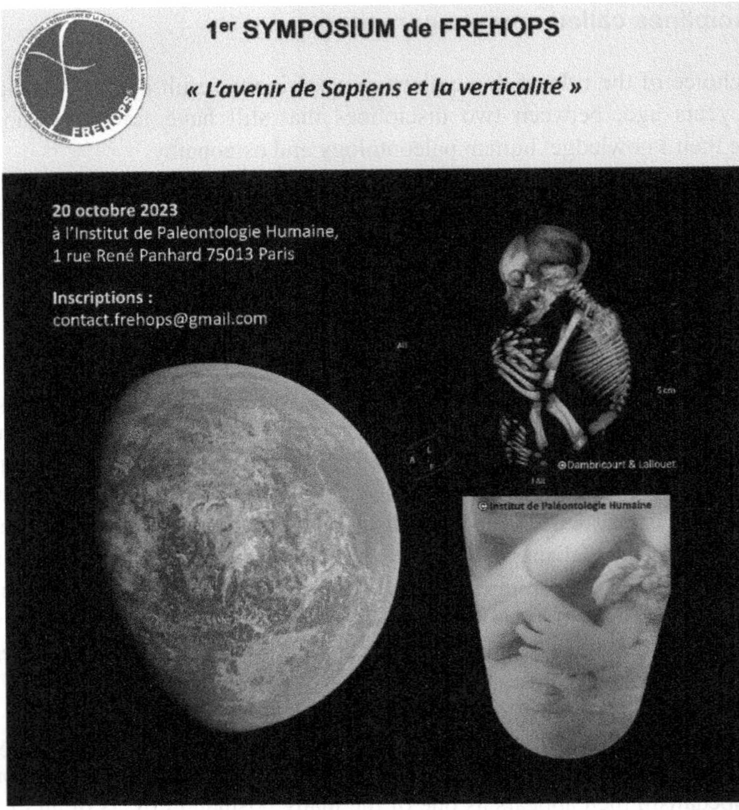

Figure P.1. *Illustration from the "The Future of Sapiens and Verticality" symposium poster. Gorilla fetus: collection of the Institute of Human Paleontology (© FREHOPS). For a color version of this figure see www.iste.co.uk/dambricourtmalasse/futureofsapiens.zip*

Famous worldwide for the discovery of "Lucy", the small bipedal *Australopithecus* whose arboricolism was met with great surprise, Yves Coppens (1931–2022) accompanied us in the creation of FREHOPS, unique in its unifying vocation and its future-oriented perspectives. He was, in this capacity, its honorary chairman. With this first symposium and this collective work, *The Future of* Sapiens *and Verticality*, we pay tribute to his memory and to his exceptional open-mindedness, following in the footsteps of Lamarck and Teilhard de Chardin (1949), who united us:

> This phenomenon is of course observable only in the long term, but it reflects a logic of astonishing stability, an irreversible time that

> memorizes! The sense of the history of inert, living, thinking matter, in its race to complexity, is a troubling observation (Coppens 1996).

The poster of the symposium echoes this change in perspective. It is inspired by Stanley Kubrick's film *A Space Odyssey* (1968), which ends with a human fetus with its eyes abnormally opened onto planet Earth, more oceanic than terrestrial, so as never to forget, as we have not ceased to recall over the last 40 years at the National Museum of Natural History, that the evolution of the animal kingdom, and ours in particular, has never left the aquatic environment, that the amniotic fluid is a preservation of this, and that our future is conditioned by the protection of our first matrix, the "planetary placenta".

February 2026

P.3. References

Coppens, Y. (1996). Grands noms, grands moments de l'histoire de nos sciences. In *Annuaire du Collège de France 1995–1996. Résumés des cours et travaux, 96e année*. Collège de France, Paris.

Lamarck, J.B. (1802). *Recherches sur l'organisation des corps vivants*. Maillard Librairie, Paris [Online]. Available at: https://gallica.bnf.fr/ark:/12148/bpt6k4226016j/f12.item.

Teilhard de Chardin, P. (1949). *La place de l'homme dans la Nature. Le groupe zoologique humain*. Albin Michel, Paris.

Acknowledgments

Anne Dambricourt Malassé; Bernard Autet, chairman of FREHOPS; Sandra Joffroy and Djillali Hadjouis wish to thank Professor Henry de Lumley, director of the Institut de Paléontologie Humaine (Institute of Human Paleontology), for making the amphitheater and reception spaces available to the symposium, as well as for his warm and informative welcome to this renowned research establishment. Offered to France by Prince Albert I of Monaco in 1910, the Institute was to enable human paleontology, for the first time in its history, to have a research center entirely dedicated to the ancestors of humanity and to understanding their biological, anatomical, cognitive, cultural and behavioral evolution, inseparable from their environments.

Introduction

Verticality, the Guiding Thread of Our Origins and Our Future

I.1. Verticality or bipedalism: a problem based on a pioneering, but inexact, thought experiment

Ever since Aristotle (384–322 BCE), the verticality of the nervous system has been the anatomical criterion that distinguishes human beings from other animals. The animal kingdom was classified into archetypes until 1802, when Jean-Baptiste de Lamarck (1744–1829) broke with more than two millennia of fixism. The acquisition of verticality emerged as the evolutionary problem of the origins of *Homo sapiens*, on which speech and the use of two hands to create manufactured objects depend, allowing the development of abstract thought with the creation of symbols, which enabled the invention of mathematics, non-existent in other species such as the chimpanzee (Radford et al. 2009).

Lamarck (1802) dared to interpret the fixist classification of present-day primates as the consequence of a natural lineage linking primates in the past, then divided into three successive groups: 1) arboreal quadruped prosimians, such as the present-day lemurs of Madagascar; 2) semi-erect quadruped and arboreal simians, with large-sized species, such as the chimpanzee and the orangutan; 3) *Homo sapiens*, erect, bimanual and bipedal with a larger brain and cerebellum than in any other primate. Lamarck was a specialist in invertebrates, however; he was not familiar with the internal organization of the axial skeleton, segmented from the posterior base of the neurocranium, down to the last coccygeal vertebra. He attempted to explain, through a thought experiment, how the morphologies of vertebral animals, delimited into distinct anatomical groups, could have transformed.

Introduction written by Anne DAMBRICOURT MALASSÉ, Bernard AUTET, Sandra JOFFROY and Djillali HADJOUIS.

He focused his thinking on the locomotor, or appendicular, skeleton, without concerning himself with the internal base of the skull, which articulates with the spine. His lack of knowledge in this area was used by anatomists to justify their opposition regarding the differences between the three groups of primates: Armand de Quatrefages (1810–1892) who had defined the Cro-Magnon race, and Paul Broca (1824–1880), who had located the articulated language area in the left prefrontal cortex. Locomotion could not account for the internal spatial organization of the skull base on which the dura mater is secured and envelopes the brain, cerebellum and brain stem as far as the coccyx (Dambricourt Malassé 2022a).

Lamarck's thought experience was appealing, and accessible to the imagination. It pleased Charles Darwin, who incorporated it in 1872, convinced of the heredity of acquired characteristics, and recognizing that the simian lineage of humans was the thesis of Lamarck, professor and chair at the Muséum national d'histoire naturelle (French national museum of natural history) in Paris, formerly the Jardin royal des plantes médicinales (royal garden of medicinal plants). Thus, despite the opposition of anatomists and physicians who had sound knowledge of humans and primates in general, locomotion eventually emerged as the mechanical cause at the origin of the verticalized anatomy of *Homo sapiens* and their fossil ancestors. The concept of natural and sexual selection adds dynamics on a larger scale, that of societies and their environments, to account for the long-term viability of transformations. The self-organization of complex interacting systems was implicit.

This lack of understanding of the internal anatomy of skeletons had the effect of shifting the question of the origins of human anatomy away from that of internal verticality, to focus on that of exclusive bipedalism. This resulted in a division of the skeleton that is still used today, with, on the one hand, the cerebral and facial skull, devoted to cognitive, language and manducatory functions, and, on the other hand, the postcranial skeleton, protecting the heart and lungs, and the skeleton's locomotor system (including the mother with her fetus), which does not allow for integration of the skull base and its synchondroses into a consistent understanding of morphological evolution, from occlusion, or dental intercuspidation, to locomotion, via the central nervous system and the sacrum.

The posterior base of the skull articulates, on its external face, with the mandible by the condyles of its ramus, as well as, particularly, on its internal face, with the spine united since embryogenesis by the dorsal chord and the membranes surrounding the neural tube. Furthermore, contrary to the long-standing assumption by Louis Bolk (1926), who focused on the position of the foramen magnum in primates, the verticality of the posterior base is not the original embryonic conformation. As we have shown, it is an acquisition due to embryonic mechanical forces that break the craniocaudal linearity of the embryo (Dambricourt Malassé 1988, summarized in Dambricourt Malassé 2022a). It begins in the center of the skull base, at the apex of the chord, at the boundary with the prechordal plate. The part of the axial skeleton that has verticalized thus corresponds to the chordal axis of the embryo, from the post-sphenoid to the coccyx. The division of the skeleton into cranium and post-cranium therefore runs counter to the natural organization of its morphogenesis, whereas it is more consistent to distinguish the axial skeleton, which protects the central nervous system, as well as the heart and lungs well before birth, and the appendicular skeleton, which will allow it to move around after birth.

Fossil species of the genus *Homo* were bipedal, so why be interested in their posture? Gibbons, small, arboreal primates existing today, are exclusively bipedal on the ground. Yet, their brain stem is poorly straightened, their cerebellum is in a high position and their bipedalism is not that of a *Homo sapiens*. There is nothing human or hominid about their posture, their occlusion or their psychomotor development. Verticality is barely inscribed in the chordal axis.

NOTE.– The traction forces of the lower appendicular skeleton straighten the spine when walking, but they do not straighten the inner posterior base of the skull. Understanding the origins of our anatomy organized around internal verticality must therefore be refocused on the orientation of the axial skeleton from the posterior base before birth, before being carried by the appendicular skeleton. This refocusing is essential in order to consider the position of the cerebellum and the brain stem, the cortico-cerebellar connections and the balance between face and base, before postnatal functions, because these are consecutive to the transformations put in place before birth. For the last 40 years, we have been insisting on the intrauterine period, with all the resulting implications for mothers and infants, from the individual to the social scale, placed back into their environmental, cultural and behavioral context (Dambricourt Malassé 2022a).

I.2. Recalling phylogenesis and the embryonic origin of axial straightening

Phylogenesis, or the classification of present-day primates and fossils, has not changed since Lamarck. It has been complemented by fossils that have found their place in one of the primate categories: prosimian, monkeys, great apes and *Homo sapiens*, with the addition of extinct groupings that geochronology places after the great apes and before *Sapiens*[1]: *Australopithecus*, *Paranthropus* and

1 The name *Sapiens* is attributed to the morphogenetic body plan of human anatomy known as "modern", or *Homo sapiens* Linnaeus, 1758. The genus *Homo* was defined by Linnaeus based on current human anatomy. It was generalized by the geneticist Ernst Mayr to extinct species, without taking into account this first level of organization of life, yet they form a different body plan. The result is a contradictory nomenclature, as our anatomy is not, in this

Homo defined by extinct anatomies. The classifications have been enriched within the groups' boundaries (more species or more genera). The introduction of morphogenesis has reinforced the observation of grouping in the basic embryonic body plan. It shows that the phylogenesis of *Sapiens*, which begins with the emergence of the first monkeys approximately 45 million years ago, is, at a minimum, the consequence of an increase in complexities in the regulation of neurulation[2] (Dambricourt Malassé 1988, 2022a; Chaline and Marchand 2010). This classification can be read in *The Cambridge Encyclopedia of Human Evolution* (Jones et al. 2009) built on adult skulls viewed externally. We simply need to split the skull in two, in the plane of symmetry, in order to observe a geometric order known to primatologists for more than two centuries and already measured by Delattre and Fenart (1960) thanks to profile teleradiographies: a gradation from a flat skull base extended by a long and narrow muzzle, and ending with a skull base and a short face, which are verticalized and wide, *Homo sapiens*. These correlations were geometrized in profile view by Gudin (Godard and Ducasse 1973) using a telecephalometric protocol in order to understand the origin of occlusion disorders in modern human populations, in relation to the orientation of the clivus and the cervical vertebrae (see Figure I.1).

A similar protocol, adding the transverse plane, made it possible to model a "basi-craniofacial contraction" recorded in the angles of the mandible (Dambricourt Malassé 1987, 1988, 1993, 2022a). It amplified from monkeys to the threshold of verticality, visible approximately 4 Ma, as of which the central nervous system could now only move around by supporting itself on the two hind limbs, which had become inferior (see Figure I.2).

This threshold was not the last: fossils show a new contraction and verticalization with *Sapiens* approximately 230,000 years ago. The curve thus shows an acceleration that reaches the "wall of internal verticality", a curve that corroborates the dazzling acceleration in the creation of manufactured tools, a creation in integrative loops that have self-complexified to the point of naturalistic and symbolic representations of the body, and even those of the self. Indeed, while more than 10 million years separated two amplifications of the straightening of the base to the threshold of its verticality, fewer than 4 million years separate the first threshold of straightening from that of *Sapiens*, ours.

This curve highlights an increasing instability in the balance of the central nervous system with a complexification of the cerebral neocortex unanimously recognized by paleoanthropologists and paleontologists.

---

case, a species of the extinct *Homo* body plan. *Sapiens* is therefore automatically elevated to the rank of genus in the Linnaean nomenclature (Dambricourt Malassé 2017).

2 Stages of neurulation. See: https://en.wikipedia.org/wiki/Neurulation.

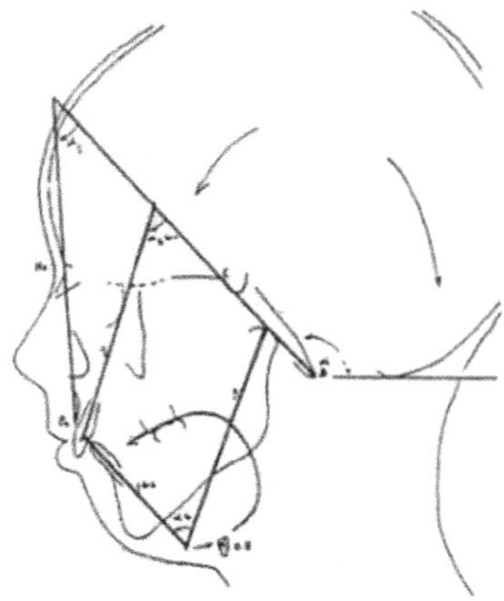

Figure I.1. *Gudin's pantograph (Godard and Ducasse 1973)*

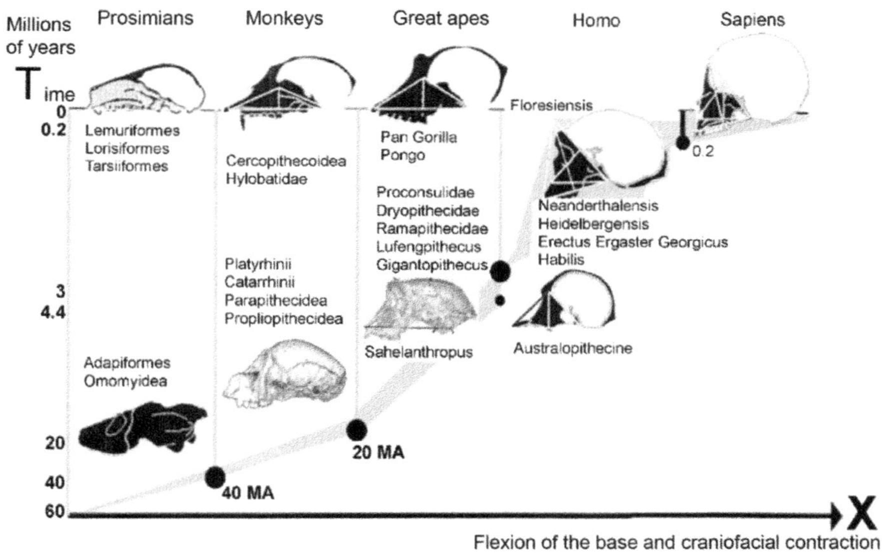

Figure I.2. *Phylogenesis of present-day primates and fossils, showing in the median sagittal view, the contraction of the adult base and mid-face visualized by Gudin's pantograph ((Chaline et al. 2010), taken from Dambricourt Malassé (1993), updated with the new fossils)*

It also shows that the hominization process, which manifests itself in the straightening of the segmented axial skeleton, has led to the increasing instability of the cerebellum, which has come to be placed above the spine and pelvis, and subsequently necessitated the complexification of the control of our own balance, or more precisely of the cerebellar lodge, involving more synaptic connections with the cerebral neocortex. Weawer (2005) has demonstrated an acceleration of cerebellization over neocorticalization over the past 10,000 years, so much so that the question arises as to whether hominization is primarily a cerebellization over cerebralization (Dambricourt Malassé in press).

I.3. A demonstrative case study: Toumaï

The heated disagreements over the interpretation of the famous fossil, *Sahelanthropus tchadensis*, nicknamed Toumaï, dating back 7 million years (Ma), with the complete skull, on the one hand, and fragments of femur and ulna, on the other hand, illustrate the consequences of failing to consider embryogenesis and its dynamics. The skull, particularly deformed at the time of its discovery, has been virtually reconstructed, but without considering the internal base, even though it is exceptionally preserved and allowed for the first time its comparison with a present-day great ape. This fossil has nevertheless been presented as the oldest hominid, i.e. a large primate, exclusively bipedal on the ground, the ancestor of the human lineage, and not as a quadrupedal, occasionally bipedal, great ape, the ancestor of gorillas, as paleoprimatologists were quick to propose (Wolpoff et al. 2002; Wood 2002). This attribution of the skull to the oldest known hominid was largely based on the position of the foramen magnum, which is assumed to have been more advanced (Zollikofer et al. 2005). Yet, the U-shaped morphology of the maxilla, associated with the large size of the root of the canine and its crown, which reaches beyond that of the other teeth, are characteristic of a great ape, and a mandible, also U-shaped, that is long and narrow, and whose condyles are in a high, receded position. They articulated with a semi-straightened internal base and a high cerebellar lodge, which is precisely the opposite of a hominid. A comparison between the skull in its original state and the virtual reconstruction confirmed the reasons for doubt, showing the greater foramen magnum in an erroneous position. It is too far forward relative to the conventional reference point given by the external auditory meatus. Part of the petrous pyramid has disappeared, which positions the anterior edge of the foramen magnum (the basion) at the level of the meatus, and not posteriorly as in all apes. The correction repositioned the basion (Dambricourt Malassé 2006, 2010, 2011, 2022a). The Toumaï skull is that of a great ape whose balance between occlusion and axial skeleton was overhanging on a semi-erect clivus–coccyx axis; the animal moved in quadrupedal mode and not in exclusively bipedal balance.

The position of the foramen magnum is determined by the degree of inclination of the clivus and the petrous pyramids that form the anterior edge of the cerebellar lodge. This degree is straightened in hominids, *Australopithecus*, *Paranthropus* and *Homo*. It was therefore appropriate to begin by verifying whether the internal conformation of the cerebellar lodge was simian (semi-straightened) or hominin (straightened). The simian organization would have been observed without any possible doubt, regardless of the degree of complexity of the neocortex, because the latter has not yet formed when optimal straightening of the clivus is reached at the end of embryogenesis. Computed tomography ultimately confirmed this: Toumaï's semicircular canals are characteristic of an arboreal, quadrupedal and occasionally bipedal ape (Le Maître 2015), which is corroborated by the appendicular skeleton. This large primate moved around by supporting itself on the joints of its hands (Meyer et al. 2023). Toumaï stood upright such as the gorilla or the chimpanzee existing today, and this posture was frequent in other fossil species of East Africa (*Orrorin*, 6 Ma; and *Ardipithecus,* 4.4 Ma), as well as of South Asia (*Lufengpithecus* of southern China, 6–7 Ma).

Such controversy generated by the reconstruction of the most complete fossil skull of the large primates of the Miocene illustrates the lack of curiosity of evolutionary paleontology for the turning point of axial straightening, even with the acquisition of internal verticality being the "common thread" of the hominization process, and its embryonic origin having been demonstrated, published and publicized for the past 30 years (Coppens 1996). The Lamarckian doxa of the locomotor mechanical cause therefore does not allow us to understand the place of the verticality of our central nervous system, of our occlusion and of the disorders of their balance in phylogenesis, nor, consequently, to better understand the mechanisms of morphogenetic and psychomotor evolution that concern us, yet we are at the limits of basi-craniosacral verticality.

I.4. Distinguishing between body plan and diversity, macroevolution and diversification

The skull bases of fossil primates are rare; in contrast, the mandible that articulates with the posterior base and obligatorily modifies with straightening is the most frequent bone fossil. However, the mandible of monkeys demonstrates a stability of angles that depend on the degree of straightening over several tens of millions of years. Current species of gorilla, chimpanzee and orangutan have the same angular values before birth and the same dental occlusion, with identical semi-erect axial architecture. All then have an aptitude for bipedalism after birth, but the central nervous system is still semi-erect and moves quadrupedally (Dambricourt Malassé 1993, 2022a). Their morphogenesis appeared approximately 20 million years ago. It is found with fossil mandibles such as those of the *Proconsul* and *Ouranopithecus* in Europe, the *Dryopithecus* and *Kenyapithecus* in Africa, the *Sivapithecus* and *Indopithecus* in India, the *Khoratpithecus* in Burma and Thailand,

and the *Lufengpithecus* and *Gigantopithecus* in China. Their bases are rare, but their mandibles "speak" for them. We will not be able to understand the origin of our verticality and that of the hominids that came before us if we do not recognize the level of organization affected by straightening. Fossils and present-day species show us that it is precisely the stability of axial organization that was disturbed during embryogenesis, in certain species of great apes, whereas others have transmitted it to present-day species.

The aptitude for occasional bipedalism is constitutive of the semi-straightened axial architecture; it is common to these fossil and current species of hominoids and has been transmitted for 20 million years. Its development is stable, while the ecosystems occupied by these different species were different, with adapted appendicular skeletons. Orangutans, who spend the majority of their existence in the treetops, have retained this ability by moving around on branches. Bipedalism is the occasional balancing of a common, stable architectural foundation for 20 million years, which became permanent and exclusive at the threshold of axial verticality approximately 4 million years ago. The occasional use of bipedalism was therefore the norm, in both Africa and Asia. There is no gradual "trend" toward internal straightening observed over the course of millions of generations.

This common architectural foundation corresponds to what biology calls the "body plan", or "bauplan", which is established during embryogenesis, along the cephalocaudal and dorsoventral axes, under the control of homeotic regulatory genes. The adult bauplan, or body plan, is a fundamental element of evolutionary biology, or *Evo-Devo*:

> The body plan concept is not the archaic, outdated, metaphysical concept that it is described to be by some researchers. Rather, it is based on modern evolutionary principles and provides a conceptual foundation for evo-devo research. Major evo-devo concepts such as constraint, evolvability, homology, canalization, integration and modularity, heterochrony, and other concepts are based on the existence of animal body plans (Willmore 2012, p. 230).

It was also appropriate to add to it the ontogenetic dimension that unfolds as of the formation of the embryo with its dynamic trajectories (Dambricourt Malassé 1988, 1993, 2022; Dambricourt Malassé, and Deshayes 1992; Dambricourt Malassé et al. 1999). Denis Duboule (2020) uses the expression *plan de base d'organisation corporel* in French, or "basic plan of bodily organization". It is thus by integrating embryogenesis that the straightened anatomy of an *Australopithecus* and that of the oldest species that defines the genus *Homo* will be understood. This "blank space on the map" of craniospinal embryogenesis, with its dynamics, applies to fossil hominids and therefore to Neanderthals.

The body plan is therefore to be distinguished from the diversity of locomotion modes of the appendicular skeleton, and paleontologists who study it are right to point out that that of the chimpanzee is not an ancestral model. This does not apply, however, to the morphogenesis of the chondro-splanchnocranium and its axial inclination. Locomotor diversity has allowed the dispersion of an axial body plan in various ecological niches, and its reproduction up to the present *Anthropocene* in gorillas or chimpanzees for the great apes, for example, or lemurs for the prosimians. The error since Lamarck has been to confuse the adaptive microevolution of the appendicular locomotor skeleton, which carries the axial skeleton, with macroevolution, that of the body plan whose embryonic mechanical forces have straightened the axial skeleton from the skull base, through a succession of amplifications up to the current threshold of verticality, *Sapiens*. The phenomenon, perceptible only over very long periods of time, is irreversible. In other words, the amplification process has been reproduced, but only since the last body plan, which is in itself a particularity that distinguishes it from microevolutions considered random and indeterminate. Irreversibility, on the one hand, and reproducibility only from the last level of organization, on the other hand, make it possible to induce an evolutionary specificity in the genome of which we are inheritors and carriers.

The search for stability is the universal law of thermodynamics and it concerns us (Bois 2010). A new embryonic straightening corresponds to a new balance state, and phylogenesis shows us that it has itself followed a stability, that of its dynamic and geometric trajectory toward greater axial instability and neurosensory complexity. What we do not understand, but which is also not sought outside of the still very rare macroevolutionary approach (Chalin et al. 2015; Queiros-Condé et al. 2022; Brissaud et al. 2023; Fleury 2024), are molecular mechanisms, or physico-chemical mechanisms, which have destabilized the hereditary memory of the *bauplan*, whereas they concern us here and now, having reached the "wall of internal verticality".

I.5. References

Bois, E. (2010). La dynamique posturale non-linéaire. Partie I : bases et fondements. *L'Anthropologie*, 114(3), 370–387.

Bolk, L. (1909). On the position and displacement of the foramen magnum in the primates. *Proc. Acad. Sci. Amst.*, 12, 362–377.

Bolk, L. (1926). *Das Problem der Menschwerdung*. G. Fischer, Jena.

Brissaud, I., Chaline, J., Queiros-Conde, D. (2023). *L'entropie créatrice. Thermodynamique fractale et quantique de l'Univers, de la Vie et des Sociétés*. Ellipses, Paris.

Chaline, J. and Marchand, D. (2010). *L'embryon, le singe et l'homme*. Ellipses, Paris.

Chaline, J., Durand, A., Dambricourt Malassé, A., David, B., Magniez-Jannin, F., Marchand, D. (2000). Were climatic changes a driving force in hominid evolution? In *Climates: Past and Present*, Hart, M.B. (ed.). Geological Society, London, 185–198.

Chaline, J., Queiros-Condé, D., Dubois, J. (2015). *Le monde des fractales*. Ellipses, Paris.

Coppens, Y. (1996). Grands noms, grands moments de l'histoire de nos sciences. In *Annuaire du Collège de France 1995–1996. Résumés des cours et travaux, 96e année*. Collège de France, Paris.

Dambricourt Malassé, A. (1987). Ontogenèses, paléontogenèses et phylogenèse du corps mandibulaire catarhinien. Nouvelle interprétation de la mécanique humanisante (théorie de la foetalisation, Bolk 1926). PhD Thesis, Muséum national d'Histoire naturelle, Paris.

Dambricourt Malassé, A. (1988). Hominisation et fœtalisation. *Comptes Rendus de l'Académie des Sciences, Séries II, Paléontologie humaine/Human Paleontology*, 307, 199–204.

Dambricourt Malassé, A. (1993). Continuity and discontinuity during hominization. *Quaternary International*, 19, 85–98.

Dambricourt Malassé, A. (1996). Nouveau regard sur l'origine de l'Homme. *La Recherche*, 286, 46–54.

Dambricourt Malassé, A. (2006). La question des origines de la lignée humaine : les modèles évolutionnistes confrontés aux fossiles, implications en orthopédie et posturologie. *Les Cahiers du Centre d'Étude de l'Ostéopathie de la Posturologie et Sciences Annexes*, 4, 38–57.

Dambricourt Malassé, A. (2010). The relationships between occlusion and posture in the hominid lineage, implications for the transition between Mesolithic and Neolithic populations. *International Journal of Modern Anthropology*, 3, 14–63.

Dambricourt Malassé, A. (2011). Les équilibres bipèdes permanents, origine embryonnaire, morphogenèse, équilibre occluso-postural, conséquences pour l'évolution psychomotrice et comportementale des hominidés. HDR Thesis, Université de Compiègne, Compiègne.

Dambricourt Malassé, A. (2022a). *Embryogeny and Phylogeny of the Human Posture 2. A New Glance at the Future of our Species*. ISTE Ltd, London and Wiley, Hoboken, NJ.

Dambricourt Malassé, A. (ed.) (2022b). Sapiens and cognition: The optimal vertical nervous system. The last threshold of self-organized and self-memorizing increasing complexity from gametes to embryo. In *Self-Organization as a New Paradigm in Evolutionary Biology*. Springer, Cham, 307–358.

Dambricourt Malassé, A. (ed.) (2022c). *Self-Organization as a New Paradigm in Evolutionary Biology*. Springer, Cham.

Dambricourt Malassé, A. (in press). Un nouveau paradigme pour une meilleure compréhension du devenir de notre espèce : le Rubicon cérébro-cérébelleux. In *Archéologie, préhistoire, paléoanthropologie, etats sanitaires. Les méthodes et les questionnements*, Hadjouis, D. (ed.). Presses Universitaires de France, Paris.

Dambricourt Malassé, A. and Deshayes, M.J. (1992). Modeling of the craniofacial architecture during ontogenesis and phylogenesis. In *The Head-Neck Sensory Motor System*, Berthoz, A., Vidal, P.P., Graf, W. (eds). Oxford University Press, New York/Oxford, 37–46.

Dambricourt Malassé, A., Deshayes, M.J., Magnies-Jannin, F., Chaline, J., Marchand, D. (1999). A solution to human paradox: Fundamental ontogenies and heterochronies. *Human Evolution*, 14(4), 277–300. doi: 10.1007/BF02438932.

Darwin, C. (1871). *The Descent of Man, and Selection in Relation to Sex*. John Murray, London.

Delattre, F. and Fenart, R. (1960). *L'hominisation du crâne*. Éditions du CNRS, Paris.

Duboule, D. (2020). Évolution des génomes et développement. *L'annuaire du Collège de France. Cours et travaux 2017–2018*, 118, 569–573.

Fleury, V. (2024). Electrical stimulation of chicken embryo development supports the inside story scenario of human development and evolution. *Scientific Reports*, 14, 7250. doi: 10.1038/s41598-024-56686-y.

Godard, B. and Ducasse, B. (1973). L'examen radiologique en O.D.F. L'analyse de Gudin. *Revue d'orthopédie dento-faciale*, 7(3), 373–392. doi: 10.1051/odf/1973029.

Jones, S., Martin, R.D., Pilbeam, D.R., Bunney, S. (2009). *The Cambridge Encyclopedia of Human Evolution*. Cambridge University Press, Cambridge.

Lamarck, J.B. (1802). *Recherches sur l'organisation des corps vivants*. Maillard Librairie, Paris [Online]. Available at: https://gallica.bnf.fr/ark:/12148/bpt6k4226016j/f12.item.

Le Maître, A. (2015). Locomotion et morphologie du labyrinthe osseux chez les hominoïdes actuels. Application à Sahelanthropus tchadensis (Hominidae, 7 Ma, Tchad). PhD Thesis, Université de Poitiers, Poitiers.

Meyer, M.R., Jung, J.P., Spear, J.K., Araiza, I.F., Galway-Witham, J., Williams, S.A. (2023). Knuckle-walking in Sahelanthropus? Locomotor inferences from the ulnae of fossil hominins and other hominoids. *Journal of Human Evolution*, 179, 103355.

Queiros-Condé, D., Chaline, J., Brissaud, I. (2022). Quantum fractal thermodynamics to describe the log-periodicity law in species evolution and human organizations. In *Self-Organization as a New Paradigm in Evolutionary Biology*, Dambricourt Malassé, A. (ed.). Springer, Cham, 291–305.

Radford, L., Demers, S., Isaias, M. (2009). *Processus d'abstraction en mathématiques*. Centre franco-ontarien de ressources pédagogiques, Ottawa.

Van Overwalle, F., Ma, Q., Haihambo, N., Bylemans, T., Catoira, B., Firouzi, M., Li, M., Pu M., Heleven, E., Baeken, C. et al. (2024). A functional atlas of the cerebellum based on NeuroSynth task coordinates. *Cerebellum*, 23(3), 993–1012. doi: 10.1007/s12311-023-01596-4.

Weawer, A.H. (2005). Reciprocal evolution of the cerebellum and neocortex in fossil humans. *Proceedings of the National Academy of Sciences of the USA*, 102(10), 3576–3580. doi: 10.1073/pnas.0500692102.

Willmore, K.E. (2012). The body plan concept and its centrality in evo-devo. *Evolutionary Developmental Biology*, 5, 219–230. doi: 10.1007/s12052-012-0424-z.

Wolpoff, M., Senut, B., Pickford, M., Hawks, J. (2002). *Sahelanthropus* or *Sahelpithecus*? *Nature*, 419, 581–582.

Wood, B.A. (2002). Hominid revelations from Chad. *Nature*, 418, 133–135.

Zollikofer, C.P.E., Ponce de León, M.S., Lieberman, D.E., Franck, G., Pilbeam, D., Andossa, L., Mackaye, H.T., Vignaud, P., Brunet, M. (2005). Virtual cranial reconstruction of *Sahelanthropus tchadensis*. *Nature*, 434(7034), 755–7599.

PART 1

Paleontological Perspectives

1

The Future of *Sapiens*: A Question of Verticality, Proof by Neanderthal Man

1.1. At the origins of straightening: the embryo, not locomotion

Strictly speaking, phylogenesis from fossil primates to current species should follow the chronology of the embryogenesis of the future skeleton, which begins with the cartilage base of the skull. Figure 1.1 shows five examples of adult bases of current primates, according to the chronological order of the appearance of their embryonic level of organization known to fossil species. The first is a prosimian (Strepsirrhine, lemur); the base is flat; its ancestors appeared in Eurasia more than 55 million years ago (Ma). A flat chondrocranium is the conformation common to all older or contemporary mammalian vertebrates of the ancestors of this lemur; the next two are small simians or monkeys, one from Central America (*Cebus*) and the other from Southeast Asia (gibbon)[1]; the cerebellar lodge is slightly lowered and its anterior face slightly straightened; their common ancestors appeared in Asia approximately 45 Ma. The fourth is a great ape, a chimpanzee, the posterior base is significantly more straightened, but not verticalized; its common ancestors with the orangutan and gorilla appeared approximately 20 Ma ago in Africa; and the fifth is *Homo sapiens* with a verticalized clivus that appeared in Africa approximately 230,000 years ago. All are quadrupeds, with the pelvis to the rear of the cerebellar lodge and the axial skeleton carried by the four limbs, except *Homo sapiens*, whose cerebellum above the pelvis is carried exclusively by the hind limbs that have become inferior. This succession is, in reality, that of embryonic body plans, the time elapsed between two reorganizations being counted in tens of millions of years.

Chapter written by Anne DAMBRICOURT MALASSÉ, Djillali HADJOUIS, Sandra JOFFROY and Bernard AUTET.

1 Gibbons, or Hylobatidae family, are not great apes (Shao et al. 2023).

Missing between the great apes and *Homo sapiens* are extinct embryogeneses, those of the first hominins and their descendants such as Neanderthals (Dambricourt Malassé 1987, 1988, 2011, 2022a, 2022b; Chaline et al. 1998; Chaline and Marchand 2010).

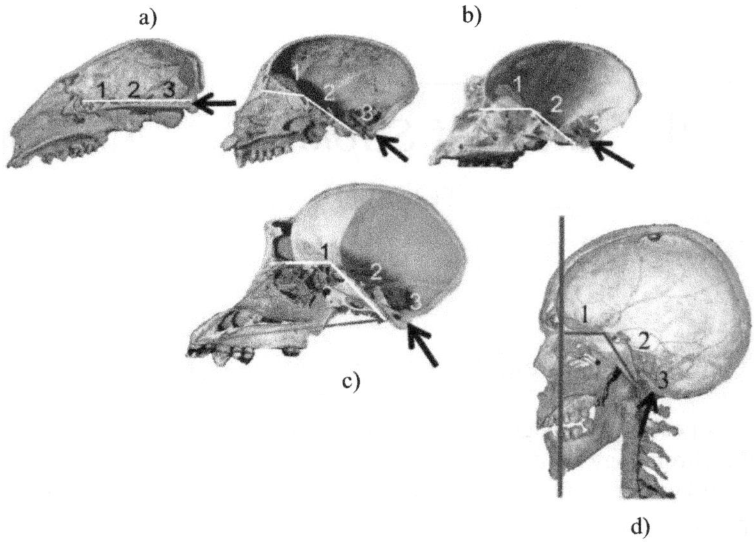

Figure 1.1. *Current species of the four embryonic body plans of primates characterized by the amplification of the straightening of the posterior base of the chondrocranium: a) prosimian; b) monkey; c) great ape; and d)* Homo sapiens *(Dambricourt Malassé 2022a). Pictures and copyright Dambricourt Malassé. For a color version of this figure see www.iste.co.uk/dambricourtmalasse/futureofsapiens.zip*

The angular shape of the human base is a dynamic acquisition of embryogenesis well before the formation of the cerebral hemispheres and thus before the vault. Guiseppe Levi (1872–1965) had already shown in 1900 that it forms flat in the human embryo, as in all vertebrates, but his illustrated descriptions went unnoticed in favor of Louis Bolk's fetalization theory (1926), which assumed an original flexed conformation common to all mammalian embryos. We unearthed Levi's pioneering works in the late 1990s, following the discovery of the embryonic mechanisms that lead to the straightening of the posterior base, verticalization of the face and exclusive bipedalism (Dambricourt Malassé 1987, 1988, 1993, 1996; Dambricourt Malassé et al. 2000a, and the Introduction to the current book). Above all, the demonstration invalidated two centuries of Lamarckian doctrine according to which axial verticalization is locomotive in origin, and updated the role of heterochronies in our understanding of morphological changes (Dambricourt Malassé 1987, 1988, 1993; Chaline et al. 1998).

Heterochronies are changes in the duration of the major phases of ontogenesis, delays, slowdowns, blockages, accelerations, extensions, for example, those of cell divisions (more neurons), or cell migrations. Cells will find themselves positioned differently, such as neural crest cells (NCCs), which run along the dorsal boundary of the neural tube, from the cephalic pole to the caudal pole, on either side of the bilateral symmetry plane. When the trajectories are rotations, they are preserved, but the angles measured using a spatial reference point have changed:

> During migration, NC cells are exposed to large number of positive and negative regulators that control where they go by generating permissive and restricted areas and by modulating their motility and directionality. In addition, as most NC cells migrate collectively, cell-cell interactions play a crucial role in polarizing the cells and interpreting external cues (Theveneau and Mayor 2012).

In addition to their high mobility, neural crest cells play a major role in shaping the pattern of the face and skull:

> The Neural crest generates species-specific pattern in the craniofacial integument, skeleton, and musculature [...]. The neural crest underlies the structural and functional integration of the craniofacial complex during evolution [...]. The role of cranial neural crest in regulating the timing of its differentiation became readily apparent (Schneider 2018).

The downward and forward rotations of the skull base reached the threshold of verticality after the embryonic amplitude of the great apes by generating new *clades*, but unlike the latter, all are extinct. These are the genera that appeared in Africa, *Australopithecus*, *Paranthropus*, *Kenyanthropus* and *Homo*, embryogeneses that are very similar, with some subtle divergences during formation of the labyrinth, for example. The locomotion of their axial skeleton was constrained by the internal verticality acquired before birth, and dural tensions, well known in osteopathy, mechanically prevented quadrupedalism. Our embryogenesis is not that of the first species that defined *Homo*; it is its reorganization, with an even more pronounced straightening. Linnaean nomenclature writes *Homo sapiens*, but embryogenesis now distinguishes between *Homo*, consisting exclusively of fossil species, and *Sapiens* (Dambricourt Malassé 2011, 2017, 2022b). When we use the binominal name *Homo sapiens*, we refer exclusively to anatomically modern humans.

1.2. Human embryogenesis

1.2.1. *Correlations of cellular movements of the neural tube and cartilage tissues of the base*

Forty years of regular syntheses on the development of the neural tube and microscopic imaging (CT scan) have allowed us to visualize the trajectories of the skull base and those of the overlying neural tube, in order to identify the mechanical forces that cause the downward and forward rotations of the central base, the post-sphenoid with the lateral otic capsules and the basi-occipital. These forces determine the positioning of the foramen magnum and the craniocervical junction. The last synthesis (see Figure 1.2) illustrates these movements and constitutes the scientific reference for the embryonic and dynamic approaches to the posture of hominids and therefore that of Neanderthals (Dambricourt Malassé 2022a, 2022b). As yet, there does not exist a more precise and documented synthesis in the scientific literature, nor are there any other geometric benchmarks for following the movements of the neural tube and the tissues of the chondro-splanchnocranium in order to observe the correlations.

Figure 1.2(a) shows the stages of the Carnegie Institution (of which there are 23 in total), from completion of closure of the neural groove, stage 13. The neural tube is shown in profile, first wound and then unwound, with the craniosacral axis in extension at the beginning of the fetal period. Figure 1.2(b) shows three stages of skull-base formation, reproduced by Levi (1900). Carnegie specifies that embryos at the same stage can be of different ages or sizes; therefore, it is the level of maturity based on multiple physical characteristics that defines a stage. Based on the size and development of Levi's chondrocrania, they correspond to stages 17 (13 mm), 18 (16 mm), 20 (17 mm) and 23 (28 mm).

The trajectories of the neural tube were followed using a given geometric reference point, the dorsal chord (x) and the axis (y) by perpendicular to its apex. The part of the neural tube overlying the chord is the rhombencephalon (future bridge, elongated marrow and cerebellum). An animation allows the movements to be followed (Paidas et al. 2001)[2]. From stage 13 (5 weeks) to stage 17 (6 weeks), the rhombencephalon elongates in front of the chord, beginning an upward rotation; this is the formation of the metencephalon. The part above the chord becomes the myelencephalon. The metencephalon extends in a generalized unwinding of the neural tube, with the spine and the spinal cord developing following this extension toward verticalization.

2 The animation shows in 60 seconds the accelerated movements of embryogenesis that follow one another in 39 days (21 days to 60 days).

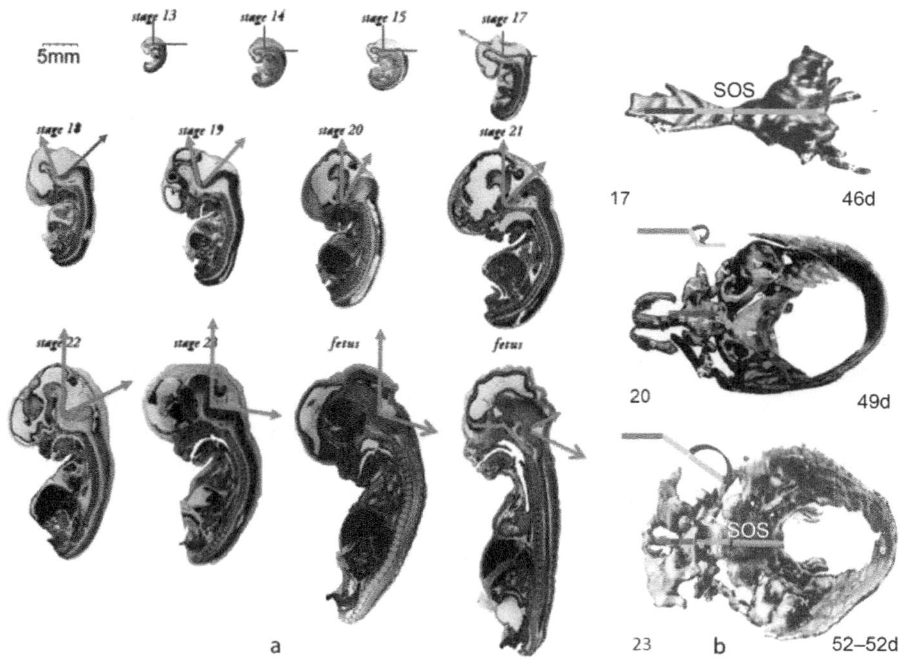

Figure 1.2. *Human embryos in sagittal and median views, and the chondrocranium in the cranial view: a) the stages of the Carnegie Institution from stage 13 to stage 23 (arrows added by the authors); b) three chondrocranial stages according to Levi (1900). The explanations are in the text (Dambricourt Malassé 2022a). For a color version of this figure see www.iste.co.uk/dambricourtmalasse/futureofsapiens.zip*

At stage 17, the base is linear, composed of cartilage condensations. Two have been able to form in the clear space in front of the chord with the ethmoid (trabecular cartilage) and the pre-sphenoid (pituitary cartilage), while the post-sphenoid and basi-occipital have formed around the chord (parachordal cartilage).

From stage 18 (7 weeks), the myelencephalon in turn begins a rotation, from bottom to top, along the y-axis. At the end of the 7th week (stage 20), its dorsal edge finds itself in contact with that of the metencephalon on this y-axis, closing the fourth ventricle. It is at this closure threshold that the post-sphenoid rotates downward and forward, pulling the median basi-occipital bone along with it without changing its rostro-caudal orientation. Levi describes a staircase profile. The otic capsules, adhering to the post-sphenoid by the cochlea, also follow the movement. The proximal end of Meckel's cartilage, which constitutes the temporary skeleton of the mandible, is in contact with the otic capsule; the mandibular cartilage and its surrounding tissues therefore follow the movement that is well illustrated in Levi. The incompressible cerebrospinal fluid, which fills the fourth ventricle before it closes, is therefore expelled into the other ventricles and the ependymal canal of the spinal cord. These internal mechanical stresses due to CSF are also to be taken into consideration.

At stage 21 (during the 8th week), a new winding elongation of the embryo causes the myelencephalon to descend past the x-axis, since the basi-occipital is lowered. In its course, it straightens it, as shown in Levi's stage 23. The post-sphenoid and basi-occipital are aligned, forming the clivus and an angle with the prechordal chondrocranium called the "spheno-ethmoidal angle". The ventral end of Meckel's cartilage, where its two sticks meet (symphyseal area), straightens. Then once more, an elongation of the spinal cord and spine, in a general unwinding motion, ends up straightening the head. The clivus aligns with the spine. Verticality is then inscribed in the morphology of the face and established along the chordal axis of the fetus from the post-sphenoid to the coccygeal vertebrae.

Using CT scans and histological sections, Utsunomiya et al. (2022) (see Figure 1.4) produced the first description of chondrification of the sphenoid (stage 17 to stage 23). While stages 17–19 are not very readable, in contrast, the difference in rotation of the sphenoid body and the alignment of the basi-occipital can be clearly seen between stages 20 and 23.

1.2.2. *Tissue forces of embryonic verticalization*

The dynamics of embryonic morphogenesis is the formation of folds, known as the "folding process". Vincent Fleury et al. (2015) have highlighted this clearly, and confirmed the following:

> In these experiments, it is assumed that the overall physical structure is the same (note: for mouse and lamprey) and that only the amplitudes of the molecular forces of development are modified. Modifying the magnitude of certain regulation forces, but not texture, leads to a deterministic trend in development. The question therefore arises as to the origin and shape of tissue texture in the early stages of head development, and whether the texture may contain a locked dynamic arrow, which would explain certain paleontological trends. In recent decades, it has been observed that physical stresses and orientation texture play a very important role in development (Fleury 2024).

The role of mechanical forces in embryonic morphogenesis is recognized but not sufficiently understood (Balasubramaniam et al. 2021). Movements are driven by microfilaments and microtubules, such as actin fiber networks coupled with myosin II (a motor protein thus forming the actomyosin complex). These networks allow the migration of cells on the extracellular matrix thanks to adhesion molecules:

> For decades, biologists have focused on the ways in which genes and other biomolecules shape bodies, mainly because the tools to analyse these signals are readily available and always improving. Mechanical forces have received much less attention [...]. As a handful of determined scientists begin to surmount those challenges, they have observed crucial forces shaping biology – from the earliest stages of an embryo's existence to diseases that strike later in life (Dance 2021).

Contemporary studies seek to distinguish between mechanical processes, on the one hand, and the control of "architect" or homeotic genes, on the other hand, specific to each homeotic compartment. This "exploration of tissue morphogenesis and homeotic gene function is essential in order to elucidate the preserved mechanisms that govern development" (Villedieu et al. 2023).

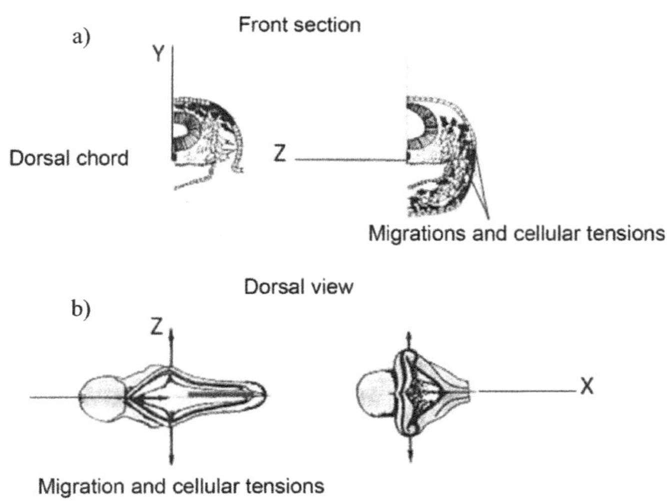

Figure 1.3. *Schematic diagram of NCC migration in the front section (a) and the force vectors closing the fourth ventricle in the top view (b) oriented in the plane parallel to the rostral end of the dorsal chord (in red), according to Langman (1984) in Dambricourt Malassé (1999). For a color version of this figure see www.iste.co.uk/dambricourtmalasse/futureofsapiens.zip*

Both views, dorsal and frontal sections of the rhombencephalon (see Figure 1.3), show a maximum amount of lateral tensions at the apex of the chord at the end of the 6th week (stage 16); there remains only the roof of the fourth ventricle (ependymal epithelium and endomeninx). An animated 3D view shows the lateral edges of the metencephalon and myelencephalon closest to the y-axis, horizontalized[3]. The chondrocranium described by Levi at this stage (16 mm) has widened considerably (see Figure 1.4(b)) while the dorsal floors and edges have continued to move closer together along the craniocaudal axis. The whole is comparable to a purse with strings that are pulled on sideways to close it (see Figure 1.3).

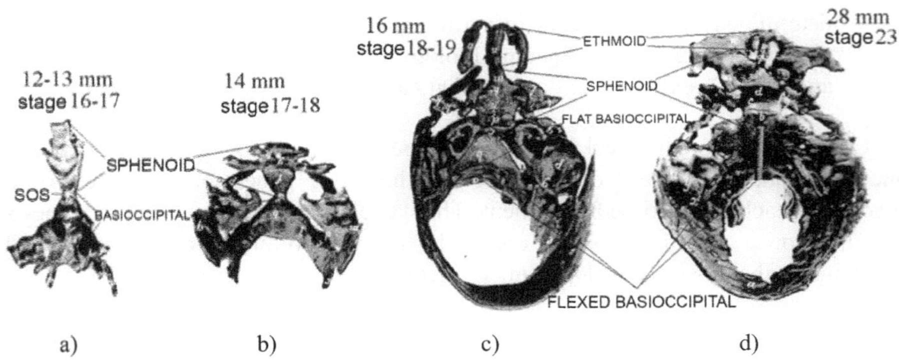

Figure 1.4. *Four stages of the human chondrocranium in the top view, described according to Levi (1900)*

The formation of the compartments corresponds to the cranio-caudal metameria of the embryo. The rhombencephalon is metamerized into eight compartments. Each forms a rhombomere (r), to which a homeotic gene corresponds. Their activation follows the rostro-caudal order and guides the migration of cells from the neural crest to the ventral pole, either by surrounding the neural tube toward the chord, or more laterally by migrating over the mesoderm that borders the neural tube. Migration begins on the 20th day (end of the 3rd week). Rotation of the post-sphenoid occurs at the end of the 7th week, and at this stage rhombencephalon r3 (HoxB2 gene) corresponds to the metencephalon and rhombencephalon r4 (HoxA1 gene) corresponds to the myelencephalon. The neural crest cells of r1 and r2 (Hox genes) fill the first pharyngeal arch (maxilla, mandible, middle-ear hammer and anvil, zygomatic bone and temporal squama). Those of r3 colonize r2 (metencephalon side) and r4 (myelencephalon side). Those of r4 fill the second pharyngeal arch

3 See: https://www.ehd.org/movies/28/The-Rapidly-Growing-Brain.

(styloid process and stapes of the middle ear). Lastly, r5 NCCs migrate into r4, and r6 NCCs migrate into the third pharyngeal arch (greater horn and caudal part of the body of the hyoid bone) (Parker et al. 2018; Méndez-Maldonado et al. 2020).

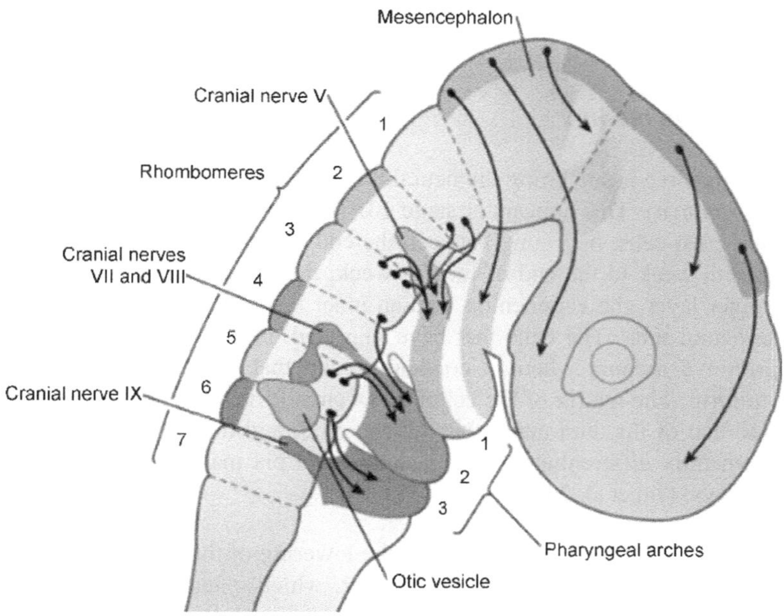

Figure 1.5. *Principal pathways of migration of cells from the neural crest of the mesencephalon to the head, and those from rhombomeres 2, 4 and 6 to the first three pharyngeal arches. The small contributions of rhombomeres 1, 3 and 5 are indicated by arrows (Morriss-Kay and Tuckett 1991). For a color version of this figure see www.iste.co.uk/dambricourtmalasse/futureofsapiens.zip*

The otic capsules are differentiated based on a superficial thickening of the ectoderm of rhombomeres 5 and 6. They are also colonized by neural crest cells (Freyer et al. 2011; Theveneau and Mayor 2012). The NCCs have therefore migrated laterally and ventrally, remaining in compartments r2 (toward the first branchial arch), r4 (toward the second arch) and r6 (toward the third arch). They have passed at the level of the post-sphenoid and basi-occipital and behind the two otic capsules (see Figure 1.5).

The tilting of the post-sphenoid and the otic capsules coincides with the maximum lateral tensions of the NCCs reaching branchial arches 2 and 3, thus passing behind the otic capsules. The identification of these forces requires consideration of the mechanical properties of the connective tissue in contact with the epithelium of the neural tube, on the one hand, and the chondrified mesenchyma of the otic capsules, on the other. These tissues are mainly linked by the extracellular matrix, which is rich in water, collagen fibers, elastic fibers and soluble molecules such as adhesion proteins.

1.2.3. *Primitive meninx*

The connective tissue lining the neural tube corresponds to the primitive meninx (*meninx primitiva*). This appears at stage 13. The meninx develops with mesoderm and neural crest cells, migrating between the end of week 3 and the end of week 5. From the 5th week to the end of the 7th week, the primitive meninx differentiates into an outer layer, the ectomeninx, and an inner layer, the endomeninx, in contact with the neural tube. The cells that have migrated under the neural tube form the basal primitive meninx where it condenses, and that is when the cartilage base begins to form. The matrix of the ectomeninx chondrifies over the course of the 7th week, and that of the dura mater that adheres to it, and forms during the 8th week. The endomeninx differentiates into arachnoid and pia mater at the end of the 8th week (57 days) (Jin et al. 2016).

The rotation of the post-sphenoid and the lowering of the otic capsules therefore precede the differentiation of the dura mater, which extends to the caudal pole, forming a dural sac around the end of the marrow (medullary cone). It is therefore secured to the straightened chondrocranium well before the fetal period. The connective spaces that allowed rotation of the post-sphenoid by means of their fluidity form the intrasphenoidal synchondrosis (ISS with the pre-sphenoid) and the spheno-occipital synchondrosis (SOS with the basi-occipital). They are composed of hyaline cartilage, rich in chondrocytes, collagen fibers and water, and are characteristic of "areas subject to significant mechanical stress" (Ferré and Salagnac 1996). By migrating into the three branchial arcs, the NCCs therefore exert tensions that interact with the adherent molecules of the extracellular matrix, on either side of the post-sphenoid and the otic capsules connected by the connective tissue rich in collagen fibers. The optimal tensions at the closure of the fourth ventricle can therefore cause them to tilt owing to the areas of least resistance, the synchondroses. The winding and unwinding phases that follow the complete straightening of the clivus probably influence the tensions of the connective tissue of the synchondroses and therefore the variability of craniosacral tensions before the fetal period. The time factor is also paramount with the reception of signals that indicate that the migration must stop as the destination has been reached. The later the signals are received, the more the amplitude of rotation of the synchondroses and the degree of straightening will increase.

1.2.4. *The holistic concepts of osteopathy*

The concepts of osteopathy have developed a holistic intelligibility of the dynamic balance of the body unknown in human paleontology and primatology, all the more difficult to integrate as the skeleton is split between the cranium and post-cranium, contrary to the unity given by the dorsal chord. This global approach would not be conceivable without the existence of fascias formed of fibrous and elastic connective tissues, which ensure the architectonic unity of the supporting tissues. They maintain the system's state of self-balance, thanks to tensegrity properties demonstrated by Guimberteau (Guimberteau 2005; Guimberteau et al. 2012; Kapandji 2012).

Moreover, osteopaths are familiar with the "craniosacral mechanism in flexion" developed by Sutherland, which maintains the longitudinal dural balance between the cerebellar tentorium, the sacrodural ligament and the terminal filum, otherwise known as the "reciprocal tension membranes" (RTM) (Hodgson 2006).

Izelfanane (2008) showed that the dural attachments are placed at the periphery of the foramen magnum, on the ventral side of the vertebral foramen of the axis and the third cervical vertebra. The dural sac is attached to the second sacral vertebra (S2) and the terminal filum to the first coccygeal vertebra. The dura mater does not transmit motion, but its flexibility allows coordination between the movements of the supporting tissues from the endocranium to the S2 canal:

> These insertions intervene in the dynamics of adaptation of the dura mater to variations in the dimension of the vertebral canal. They also constitute a means of modulating, curbing and controlling mobility and growth. With each movement of the vertebral column in flexion or extension, these attachments stretch or relax and have a veritable vascular pumping role, allowing optimum circulation between the intra- and extraspinal vascular network (Izelfanane 2011).

Remaining to be found was a unification between the dura mater and the external body, especially at the cervicocephalic junction, whose orientation is imposed before birth by connective tissue tensions in the basi-cranial synchondroses, and as they determine the position of the petrous pyramids and the foramen magnum. The only contact of the *dura mater* is with its inner layer at the craniocervical junction, via the myodural bridge. This fascia, discovered in 1995, is placed between the atlas and the occipital squama. It connects the dura mater to the suboccipital muscles, which insert onto the atlas and axis, and is observed in a wide range of mammals (Zheng et al. 2017, 2020).

1.3. Current non-human primates

1.3.1. *Embryogenesis*

The migration of NCCs to the branchial arches, and therefore under the chondrocranium, is common to all primate embryos. Studies are rare because they require embryo collections in comparative anatomy laboratories. None exist for great apes; it is necessary to refer to the fetuses.

The recent histological description of the neonatal synchondroses of prosimians and monkeys therefore serves as a reference. The study illustrates cross-sections of seven Strepsirrhines and six monkeys (Smith et al. 2021). The synchondroses are clearly visible, but none of the prosimians show angulation. In contrast, monkeys all have a spheno-ethmoidal angle, but no species has the amplitude of *Homo sapiens*. This study confirms the difference between the chondrocrania of these three major groups of primates.

In order to understand base angulation and its amplification, the regulation of homeotic genes needs to be taken into consideration. But, it is necessary to add the duration of their activity that positions the NCCs in their final destination. Duration is paramount in the coordination of rostro-caudal and dorso-ventral morphogenesis:

> Although the initial patterns of NCC segmentation and migration are conserved between species, the variety of vertebrate facial morphologies that exist indicates that a complex interplay occurs between intrinsic genetic NCC programs and extrinsic environmental signals during morphogenesis [...]. Recent findings support the view that heterochronic (in time), heterotopic (in place) or quantitative changes in the expression of key signalling molecules in the face epithelia and/or NCCs might underlie the evolution and variation of facial morphology (Minoux and Rijli 2010).

We come across the observation of “specific pattern”, “bauplan”, basic body organization plan (Duboule 2020), fundamental ontogenesis (see the Introduction to the current book) and that of heterochronies. The increase in the amplitude of rotation of the sphenoid observed during a change in the body plan can be interpreted as the consequence of the prolongation of the trajectory of the folds generated by cell migrations. They are not linear trajectories; they are curved. The tilting of the post-sphenoid, the connected otic capsules and the cartilaginous model of the mandible can be interpreted as a passive movement due to the tensions of the NCCs during their migration toward the ventral pole, generalized to the connective tissue, which ensures the balance of mechanical forces.

1.3.2. *Dynamic relationships between the skull base, occlusion and posture of great apes*

The application of basi-craniofacial angular measurement protocols to the growth of a genus of Panina (chimpanzee) made it possible to correlate the trajectories of the base in the sagittal plane with those of the mandibular symphysis and the occlusion of the canine. Figure 1.6 shows these correlations in a current great ape and an *Australopithecus* (Dambricourt Malassé 2022b).

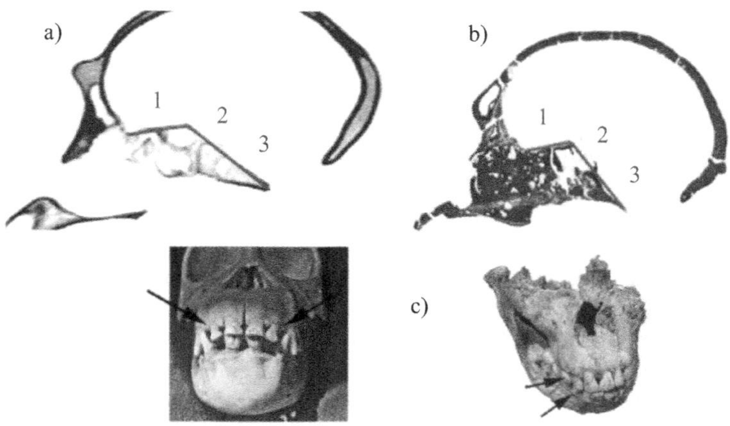

Figure 1.6. *CT scan in the median sagittal section of adult skulls of chimpanzee (a) and* Australopithecus *(Sterkfontein 5) (b). c) Canine occlusion with locking in the ape and disappearance with the hominid with a straightened chordal axial skeleton. 1: tangent to the spheno-ethmoidal plane; 2: sphenoid angle; 3: tangent to the clivus (Dambricourt Malassé 2022b). Pictures and copyright Dambricourt Malassé. For a color version of this figure see www.iste.co.uk/dambricourtmalasse/futureofsapiens.zip*

The clivus straightening angle in the great ape is significantly more open at birth than in *Homo sapiens*, the cerebellum is in a high position, and dental occlusion at the deciduous stage shows a locking between the canines and the first milk molars (dm1). After the eruption of the first permanent molar (M1), the angle of the base opens, the cerebellar lodge rises, the angle of the mandibular symphysis opens with the growth of the alveolo-dental arch forward, facial prognathism becomes more pronounced and the outline of Gudin's pantograph lowers and lengthens (see the Introduction to the current book). As for occlusion, the lower canine is in front of the upper canine. The dimensions of their cusp and of the first molar are such that they block lateral and antero-posterior movements of the mandible. Locking accentuates with the growth of the permanent fangs, which are proportionally larger. The friction of the upper canine on the caniniform lower premolar (P3) leaves the trace of a characteristic wear facet. The locking of the adult dentition is all the more pronounced when the pantograph is open, which corresponds to the loss of axial verticality and the frequent use of quadrupedalism (Dambricourt Malassé 2010).

Paleontologists have focused their attention on the C-P3 honing complex in great apes, as masticatory function is associated with the development of powerful masticatory muscles but without establishing links with the balance of forces between the face and the overhanging base on the cervical spine.

It is very likely that during the dorsal ascent of the cerebellar lodge and the receding of the temporo-mandibular joints, the myodural bridge plays a role in balancing tensions with the fascias of the neck muscles, which insert onto the atlas and axis. They exert tensions transmitted by the dura mater to the sphenoid at the periphery of the cerebellar lodge, itself stretched to the sacrum.

Comparisons with *Homo sapiens* are made using homologous characteristics and angles. However, the angle of the adult ape base is not representative of embryonic rotation since the cerebellar lodge moves upward after birth. A new angle has therefore been added to the classical protocols, the "sphenoidal rotation" angle at stage 20 of embryogenesis, which takes into account the trace of the SOS on the pharyngeal surface of the clivus, rather than the anterior edge of the foramen magnum (basion) (Dambricourt Malassé 2005). The protocol shows that this angle is stable in the great ape and confirms that the embryonic rotation is of lesser amplitude (Dambricourt Malassé et al. 2025), as observed on a fetus and newborn of the Panina (Gorilla) (Dambricourt Malassé et al. 2000b).

Different craniofacial angular measurement protocols were applied to a series of adult skulls of *Homo sapiens,* and to fossil species of *Homo* and *Australopithecus* (Dambricourt Malassé et al. 2000, 2025; Dambricourt Malassé 2002, 2022b). The results show that all hominins have a significantly closer base straightening angle than a great ape, and that the cerebellum lodge does not rise, axial verticality is maintained, quadrupedalism has disappeared and bipedal balance has become permanent and more unstable for the cerebellum. The dentition no longer shows any fangs or caniniform first molars; the mandibular canine of the *Australopithecus* is still in front of the maxillary canine, but it is small and very rarely, and only slightly, protrudes beyond the occlusal surface of the other teeth. The deciduous molars of the mandible form five cusps such as the molars, rather than two cusps as in great apes. The morphogenesis of the tooth germs took longer to form the cusps. The anteroposterior and lateral blocking of occlusion has disappeared, which is consistent with the disappearance of craniocaudal tensions, in quadrupedal posture, exerted on the posterior base of the neurocranium and on the temporomandibular joints (TMJ). The fossil species of *Homo* have an incisal end-to-end occlusion.

NOTE.– It is now clear to us that neural crest cells have migrated for longer and that the regulation of homeotic genes along the rostro-caudal axis of the embryo has changed. The heterochronies manifested as evidenced by the five deciduous cusps instead of two, the delay in the formation of the supraorbital torus, and a lower, and therefore more unstable, cerebellum. This unprecedented instability necessarily generated a complexification of the neural networks connected to the cerebellum so that the latter is more rapidly informed of its position (that of the cerebellar lodge and the labyrinth) and avoids falling, with the body moreover no longer supporting itself on its four limbs, but only on two.

The angles of straightening nevertheless differ between *Australopithecus* (the least straightened), *Paranthropus* (the sphenoid is more straightened than *Sapiens*, but this genus became extinct approximately 600,000 years ago), and all fossil species of the genus *Homo* are straighter than *Australopithecus*, but less so than *Sapiens*. Differences are found with the labyrinth enclosed within the otic capsule, in particular the angle between the edges of two semicircular canals, the lateral (or horizontal) and the posterior. This angle is closed in great apes and open in *Homo* and *Sapiens*. A priori, the more-open angle would have been the consequence of the greater flexion of the base and caused, according to doctrine, by bipedalism. The open angle should therefore be common to all hominins in exclusive bipedal balance on the ground. Fossils have disproved this reasoning, because the *Australopithecus* canals such as Sterkfontein 5 (Sts 5) in South Africa are simian:

> The overall shape and the semicircular canal size of the labyrinths of Sts 5 and other specimens of *Australopithecus africanus* are closest to the ancestral morphology shown by the great apes. In contrast, the labyrinth of *Homo erectus* was found to be close to that of modern humans. That all great ape species, as well as the 10-Myr-old hominoid *Dryopithecus,* share the same basic morphological pattern suggests that this represents the ancestral human condition (Spoor and Zonneveld 1997).

We compared a five-month-old gorilla fetus in utero with two human fetuses of the same age, one normal and one anencephalic without rotation of the skull base, to verify whether the angle between the two canals was absent in the latter fetus. Our assumption was invalidated: the otic capsules did not pivot, but the canals formed normally; their morphogenesis is therefore independent of the position of the otic capsules and the rotation of the sphenoid (Dambricourt Malassé and Lallouet 2009). The proof was thus given, *contrary* to the doctrine in force, that the transformations at the origin of the straightened craniosacral morphogenesis of the *Australopithecus* were not due to bipedal locomotor stresses. They are embryonic, in continuity with the lineages of primate species that have transmitted processes of self-organization of embryogenesis, significantly constrained by homeotic genes (Dambricourt

Malassé 2011, 2022a, 2022b). Research into embryonic morphogenesis of the labyrinth had already found that cartilage condensations of the semicircular canals underwent apoptosis (Fekete et al. 1997; Nishikori et al. 1999) under the control of these genes (homeobox Hmx) (Wang et al. 2001).

Changes as significant as the orientation of the chordal axis of the embryo and the morphology of the labyrinth therefore preceded birth. That of the labyrinths no longer changes. These changes have impacted posture and its control, but unlike classical thought patterns, it was not the posture, the locomotion type or the size of the brain and the cerebellum that generated them after birth. They are modifications in the regulation of embryogenesis according to the rostro-caudal gradient, which resulted in a prolongation of the earliest phases with heterochronic and postural consequences. The complexity in the coordination of homeotic genes has extended over time.

1.4. Occluso-postural balance of *Homo sapiens*: the contribution of dento-facial orthopedics and the examples studied

The clinical observation of balance relationships between occlusion, the orientation of the mandibular symphysis, the clivus and curvatures of the spine was established in the 1950s by Robert Gudin, thanks to the use of profile teleradiography and the application of an outline, called a "pantograph". In 1989, Gudin and Godard concluded that "tongue-mandible-hyoid balance depends on a regional vertebral-craniofacial system and on the balance of the whole vertebral column" (Gudin and Godard 1989, Figures 9 and 10). At the same time, a DFO collaboration integrated the first comparisons with great apes, helping to refine understanding of the relationships between malocclusions, the dynamics of synchondroses and vertical posture (Dambricourt Malassé and Deshayes 1989, 1992; Deshayes et al. 1990; Deshayes and Dambricourt Malassé 1991). The clinical success of orthodontic treatments in temporary dentition is obtained with the application of forces to the vault and the dental arches through the use of orthopedic masks (dental class II: high occipital squama, mandibular incisor–canine row too recessed; dental class III: occipital squama too low, row in front of its maxillary antagonist) (modified Delaire cephalometric analysis, Deshayes 1986). Treatments have proven effective before the age of 6, with a return to class I showing the closure or opening of the sphenoidal angle depending on the direction of the applied forces, activating flexion or slowing it down (Deshayes 2006). Chewing and good posture then help to maintain the restored balance between the tensions of the different mesenchymal matrices present (Deshayes and Deshayes 2013). These results confirm the dynamic unity of fascias from the exoskeleton to the basi-cranial synchondroses.

1.4.1. *Contemporary sapiens: occluso-postural imbalances*

Posturology, or the science of posture, was defined by Pierre-Marie Gagey in 1980 to understand the mechanisms of maintaining an upright position, and its stability, and to constitute an etiology of dysfunctions of the standing position (Gagey and Weber 2005). In 1984, Bernard Autet developed a technique to test the tone of the external rotator muscles of the hips in decubitus, considering a multimodality of "postural" inputs, constituting the "postural system": the locomotion of the lower appendicular skeleton would interfere via the fascias with the TMJ and the occlusion (Autet 1985). In 1989, Clauzade and Daraillans insisted on the relationship between occlusion and posture, and developed the osteopathic concept of occlusion, focusing on the trigeminal nerve (V) (Clauzade et al. 1989; Clauzade and Daraillans 1992; Clauzade et al. 2016). All these clinical observations confirm the orthodontic premises of Gudin and his colleagues (Gudin and Godard 1964; Gudin and Ferré 1968).

A recent study verified the influence of the Autet test on occlusion in children in mandibular rest, in maximum intercuspidation and without dento-dental contact. The results show that variations in occlusion induce a modification of the test generally in the direction of isotony (Conde Vázquez 2019). Nevertheless, fascias alone are not sufficient to account for facial movements under the control of the trigeminal nerve; a neural connection with the CNS is necessary (Conde Vázquez and Suárez-Quintanilla 2020). In 2023, a study of stabilometers concluded that the links between the stomatognathic system and posture were real:

> Both orthognathic surgery and orthodontic braces significantly improve postural balance and athletic performance. Additionally, there are postural differences between Angle occlusal classes, with Angle classes II and III causing changes in postural values. Some studies also suggest that dental occlusion may influence posture, but only under specific movement, fatigue, and visual deprivation conditions. Occlusal devices, braces, silicone plates, and mouthguards, designed to simulate a malocclusion, impact the patients' postural system response (Ionită et al. 2023).

The cerebellum is central to the coordination of these involuntary, and voluntary, movements. It receives nerve influxes through the neurosensory receptors of the proprioceptive system, superficial (cutaneous) and deep (musculotendinous and articular), via the pons and the spinal cord. The maintenance of the head's balance is enabled by the eyes and the vestibular system (semicircular canals and cochlea), as well as by sensors such as the TMJ and the occlusion, which are themselves sensitive to craniosacral tensions.

1.4.2. *Loss of incisal end-to-end occlusion over the course of evolution*

Sapiens appeared approximately 300,000 years ago in Africa, with no change in end-to-end occlusion for nearly 270,000 years. Indeed, the overlapping of the upper incisors, or psalidontia, is not observed before 30,000 years in North Africa, with the populations of Afalou. The study of malocclusions in contemporary as well as historical populations of the Paris Basin benefits from a new understanding thanks to their comparison with this change in dental articulation in populations of the Upper Paleolithic, as well as the Neolithic.

As we have just recalled, the dental articulation of hominins plays a key role in the dynamic behavior of the maxillofacial sphere and its functions are intimately linked to the whole of the basi-craniofacial architecture. The establishment of psalidontia is only morphogenetically affirmed with the new *Sapiens* body plan, but not before 30,000 years in North Africa. The oldest *Sapiens*, said to be robust due to their exocranial superstructures, such as the Levantine populations represented at Skhul (140 ka) and Qafzeh (95 ka), show that they had retained end-to-end occlusion (labidontia). The new configuration and function of the *Sapiens* dental articulation in psalidontia therefore appeared very late, after the architectonic transformations of the neurocranium, the basicranium and the three layers of the *Sapiens* face.

The works undertaken on the Paleolithic populations of the Upper Pleistocene, North Africa and the Sahara, at Afalou Bou-Rhummel (Hadjouis 2002, 2003, 2011), Taforalt (Aoudia-Chouakri 2003; Hadjouis 2015), Columnata (Chamla 1970; Hadjouis 2015), and Taza 1 (Hadjouis et al. 2000a) and Neolithic populations, at Tin Hanakaten (Heim et al. 2011) and Oued Guettara (Hadjouis et al. 2000b), showed a new craniofacial and occlusal anatomical reading by studying the repercussions on the face and/or on occlusion when one of the parameters of this dynamic is lacking. These populations of anatomically modern humans presented for the first time two major facts:

– The psalidontal dental articulation is represented in *Homo sapiens* from the Upper Paleolithic (hunter-gatherer lifestyle) and not the Neolithic (animal domestication and agriculture).

– Psalidontia is present in the young (with mixed or permanent dentition) and in adults.

This last point was studied rigorously by Petit-Maire on Neolithic populations (Holocene) of the Atlantic Sahara and Mauritania (Izriten 6,100 ± 120 BP, Sebkha d'Amtal between 3,000 and 2,000 BP, Sebkha Maharia 5,040 BP and 4,100 BP, and Tintan and Chami between 3,580 and 2,100 BP). The anthropologist and her collaborators suggested for the first time, on the basis of these Saharan Neolithic populations, that the passage from a labidontal to psalidontal articulation would be due to the strong dental abrasion in adults, causing an end-to-end occlusion (Petit-Maire and Charon 1979; Petit-Maire and Fleitcher 1979; Petit-Maire et al. 1979).

This functional interpretation is supported by the populations of the Upper Paleolithic sites studied – Afalou, whose skulls and teeth of the juvenile series are perfectly preserved (Afalou 4, Afalou 6 and Afalou 26) (Hadjouis 2011, 2015), and Columnata (H21 and H43) (Chamla 1970; Hadjouis 2002, 2011) – and the Saharian Neolithic site of Tin-Hanakaten (child no. 5) (Heim et al. 2011) because all are in psalidontia and no overlap of the upper incisors is altered by an abnormality or early basi-craniofacial dysmorphosis.

Accustomed to a ritual, but not widespread, practice, the populations of Afalou, Taza 1 and Oued Guettara experienced dental avulsion, which consists of removing the upper incisors, sometimes with the canines, in individuals of both sexes during their adolescence. The result in adults is a hypomaxillia (growth insufficiency) at the mid-face, while at the lower level, the incisors and canines have experienced excessive development in height with an absence of the lingual guide that normally rests on the central incisors and contributes to its longitudinal growth. In adult individuals not affected by this ritual, the severe dental abrasion gradually shifted the dental articulation toward incisal end-to-end (psalidontia/labidontia or total labidontia), according to the physiological mechanism of mastication similar to that noted by Petit-Maire in adult populations of the Atlantic Sahara. It is therefore not a question of class III as observed in the current patient base.

This observation of inversion of the dental articulation during the life of the individual (from psalidontal to labidontal) is observed in the historical and contemporary populations (from the Low Empire to the present) of Southeast Paris, but they are generally affected by class-III occlusal shifts (promandibulia), and bruxism. While children with healthy oral-dental hygiene and occlusion without major shifts are all in psalidontia, basi-craniofacial dysmorphoses, coupled with asymmetries, have, however, affected considerable numbers of individuals, also visible in adults, given the lack of care during the periods of growth concerned by these imbalances. In addition to the new psalidontal occlusal situation specific to the degree of *Sapiens* flexion acquired approximately 30,000 years ago, new dento-skeletal imbalances came to be added after the Neolithic, with canine locking having disappeared since the *Homo* flexion stage. Among the 10 or so medieval and modern necropoles studied in Val-de-Marne (Hadjouis 1999, 2000a, 2000b, 2017, 2018), in La Queue-en-Brie, we note a significant prevalence of class-III dysmorphosis, most often soliciting promandibulia, with brachycrania of 37%, followed by hyperbrachycephalic (12%) and ultrabrachycephalic (8%) cranial forms (Cho 2005; Cho and Hadjouis 2005; Hadjouis 2017). Excessive basicranial flexion in this population resulted in an incisal end-to-end occlusion.

These data are important for placing the current high frequency of cases of class II without it being possible to establish as complete an etiology as possible, insofar as the hominization process, when it is taken into consideration, is reduced to the reduction of the dental formula outside of occluso-postural and embryonic considerations.

1.5. The skeleton of *Homo neanderthalensis*

The study of the occluso-postural balance of Neanderthals is in its early stages. Thanks to the generalization of scanning over the last 10 years, it is becoming possible to gather conclusions relating to the different parts of the skeleton, whereas these cannot form part of global knowledge of craniosacral, internal and external stresses. However, there is a convergence toward what we have been supporting for more than 30 years:

1) the reality of a body organization plan unit established with embryogenesis;

2) the divergence of embryogenesis of Neanderthals since that of the primitive (or plesiomorphic) species, which defines the genus *Homo* (*Homo habilis*) in the sense of a basi-craniofacial extension, but within the limits of embryonic chord straightening;

3) the embryogenesis of *Sapiens* corresponds to a new threshold of embryonic rotation, a new chordal verticalization and a new craniosacral morphogenetic body plan that reaches the "wall of internal verticality".

The Neanderthal skeleton from La Chapelle-aux-Saints (45,000 years) was reconstructed by Marcellin Boule (1911) with a vaulted posture:

> An enormous head on a short, thick trunk, with short, stocky, very robust limbs. The proportions of the limbs are entirely consistent with modern human dispositions. The distal segments of these limbs were short compared to the proximal segments. The leg was, relative to the thigh, shorter than in any modern race. [...] The morphology of the ribs indicates the existence of very powerful intercostal muscles; a similar conformation is only found in carnivores (Boule 1911).

However, for the cervical spine:

> The direction of the spinous processes and the low obliquity of the articular surfaces of the zygapophyses, seem otherwise to indicate a total absence of curvature, at least a curvature less pronounced than in present humans, or perhaps even a slight curvature in the opposite direction, in a regular continuation of the dorsal curvature, as in Anthropoids and notably in chimpanzees (Boule 1911).

The skeleton shows a near absence of cervical lordosis as in the great ape, a high and dorsally extended cerebellar lodge, a prognathic face distant from the vocal tract, a horizontal orientation of the spinous processes of the cervical vertebrae and the first dorsal vertebra, a very wide thorax in the upper region, and curved femurs. This skeletal anatomy is consistent, except for the retroverted orientation of the coxal bone (see Figure 1.7).

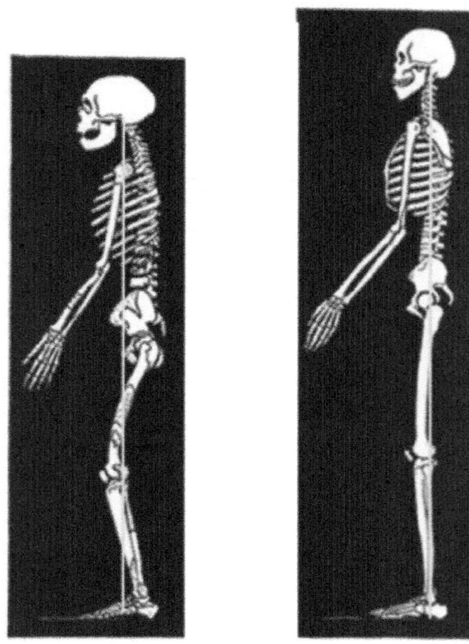

Figure 1.7. *Comparison of diagrams of the skeleton from La Chapelle-aux-Saints and that of* Homo sapiens, *each viewed in profile, according to Boule (1911). The gravitational line in yellow has been added, passing through the porion (upper edge of the auditory meatus). For a color version of this figure see www.iste.co.uk/dambricourtmalasse/futureofsapiens.zip*

1.5.1. *The head: occlusal-postural balance*

The La Chapelle-aux-Saints skull was teleradiographed according to DFO standards and its basi-craniofacial balance compared to the skull of Cro-Magnon 1 (25,000 years) by Deshayes's cephalometric analysis (Dambricourt and Deshayes 1995) (see Figure 1.8).

More recently, angular measurement protocols, including the angle of sphenoidal rotation, have been applied to scans of Neanderthals and *Homo sapiens* from the Late Pleistocene (La Chapelle-aux-Saints, La Ferrassie 1, La Quina H5, the Pech de l'Azé child, Cro-Magnon 1 and Abri Pataud), of *Homo erectus* from Southeast Asia from the Middle Pleistocene (Ngandong 7, Ngandong 12, Ngawi 1), of young and adult chimpanzees (*n* = 41), and, lastly, of adolescent and adult patients (*n* = 55) (Dambricourt Malassé et al. 2025).

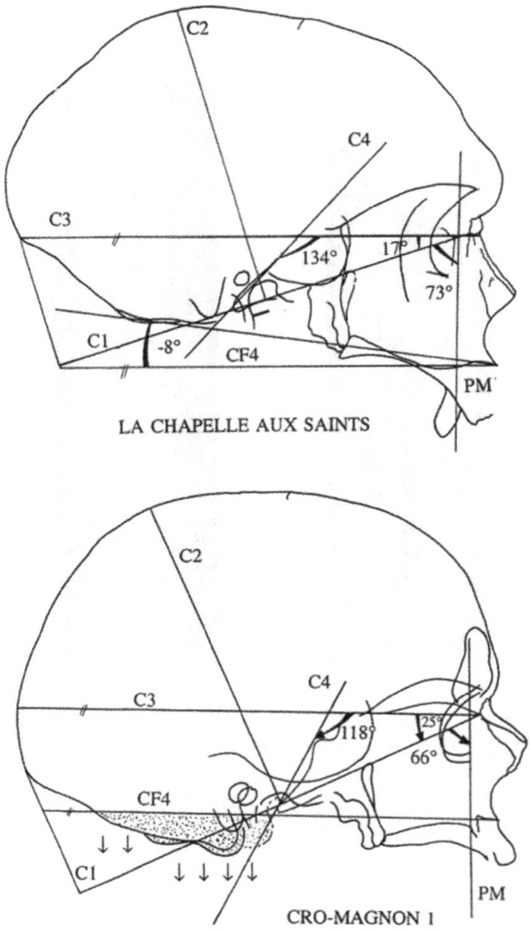

Figure 1.8. *Analysis of the craniopalatine balance of La Chapelle-aux-Saints compared to Cro-Magnon, according to Deshayes's protocol*

COMMENT ON FIGURE 1.8.– the position of line C1 relative to CF4 (orientation of the roof of the palate) highlights a different balance between the base and the mid-face. The sphenoidal angle is very open (134°) and the cerebellar lodge is in a high position (Dambricourt Malassé and Deshayes 1995).

Homo erectus and Neanderthals have less-straightened petrous pyramids than *Sapiens* (more than 90° for ours), consistent with a less-verticalized clivus slope. The slopes are even less inclined in Neanderthals. These differences were known, but the correlations between the angles, particularly the angle of sphenoidal rotation, had not been demonstrated. The angle of embryonic rotation of the sphenoid is less pronounced than in *Sapiens*.

Cephalometric analyses showed that the Neanderthal had no craniopalatine balance, unlike *Homo sapiens* (alignment of the palate, the top of the odontoid process and the lower edge of the occipital squama) (Deshayes 1991), and that the cerebellar lodge is too high. The dental drawer developed in a forward direction, leaving what is referred to as a "retromolar" space between the last tooth and the rising ramus of the mandible, while the mandibular symphysis, which has no chin, was forward inclined with the dentition. The occlusion of the canines and incisors was end-to-end, as in extinct species of the genus *Homo*.

The comparison between a skull of *Homo habilis* (KNM-ER 1813, 1.9 Ma) (see Figure 1.9) and La Chapelle-aux-Saints is illustrative. The Neanderthal is in extension, the pantograph is "collapsed", the virtual projection of the floor of the nasal cavities faces the slope of the clivus and crosses the occipital squama passing above the foramen magnum, whereas the floor of the "plesiomorphic" species is below the clivus and the occipital squama.

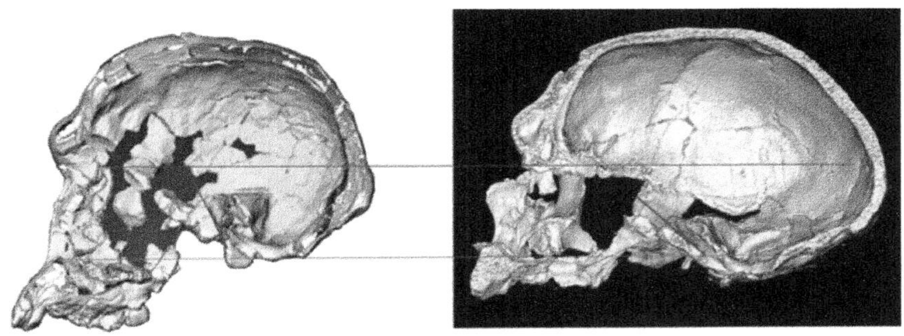

a) b)

Figure 1.9. *a)* Homo habilis. *b) La Chapelle-aux-Saints. Comparison of angles of rotation of the posterior base with visualization of the extension of the face and the occipital pole ((a) CT scan provided for teaching at the Muséum national d'histoire naturelle; (b) picture by Dambricourt Malassé). For a color version of this figure see www.iste.co.uk/dambricourtmalasse/futureofsapiens.zip*

We can already infer embryonic interpretations. The rotation of the sphenoid did not exceed the amplitude of the oldest *Homo* species. As seen previously, cephalocaudal embryonic tensions vary between the winding and unwinding of the rostro-caudal axis of the embryo, and the decreases in tension in the synchondroses are synchronous with their relaxation to the dural sac. A decrease in the tensions of the embryonic fascias can therefore be at the origin of a generalized "sagging" of the skeleton, from the face to the sacrum. Moreover, the labyrinth of Neanderthals attests to a modification in the regulation of homeotic genes upstream of its morphogenesis. Indeed, the position of the posterior canal is lowered relative to that of the lateral (or horizontal) canal (Spoor et al. 2003). The myodural bridge and the tensegrity properties of connective tissues certainly acted in the balancing of internal and external cerebrospinal tensions before and after birth.

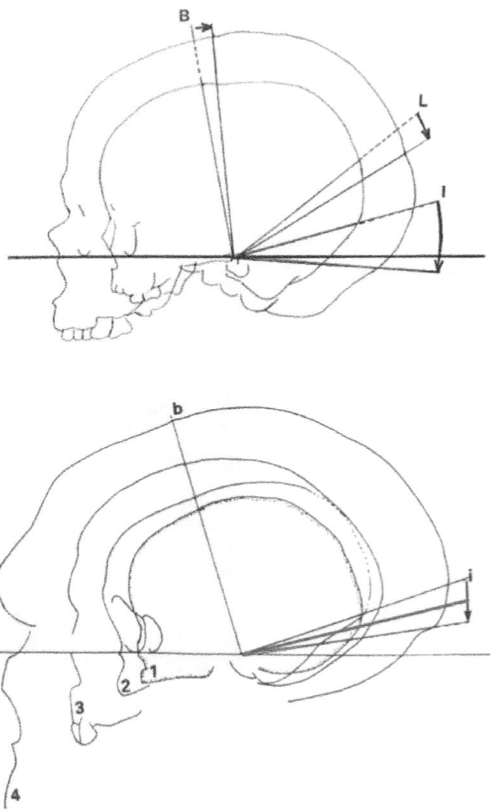

Figure 1.10. *Comparison of vault growth vectors in the Frankfurt plane between* Homo sapiens *and Neanderthal (2.3,4). B, b: bregma; L: lambda; I: inion. The green line is the position of the inion of the* Sapiens *child (1). For a color version of this figure see www.iste.co.uk/dambricourtmalasse/futureofsapiens.zip*

1.5.2. *Craniospinal balance: absence of lordosis and knees in flexion*

The fetal development of the Neanderthal occurred in a craniosacral context in extension relative to the ancestral species. The face, more prognathic than *Sapiens* at birth, will continue its development in this context, with hyperpneumatization of the maxillary sinuses (Heim 1989).

The comparison of growth trajectories with *Sapiens* (see Figure 1.10) shows the extension and very low growth of the cerebellar lodge downward and forward, and it occurs dorsally. The tensions of the synchondroses of the posterior base of the skull did not allow the winding of the encephalon, despite a brain at least as, if not more, voluminous. The cerebellum developed in an extension context, giving the neck its "bomb" shape with its occipital bun, and not a rounded profile as in *Sapiens*.

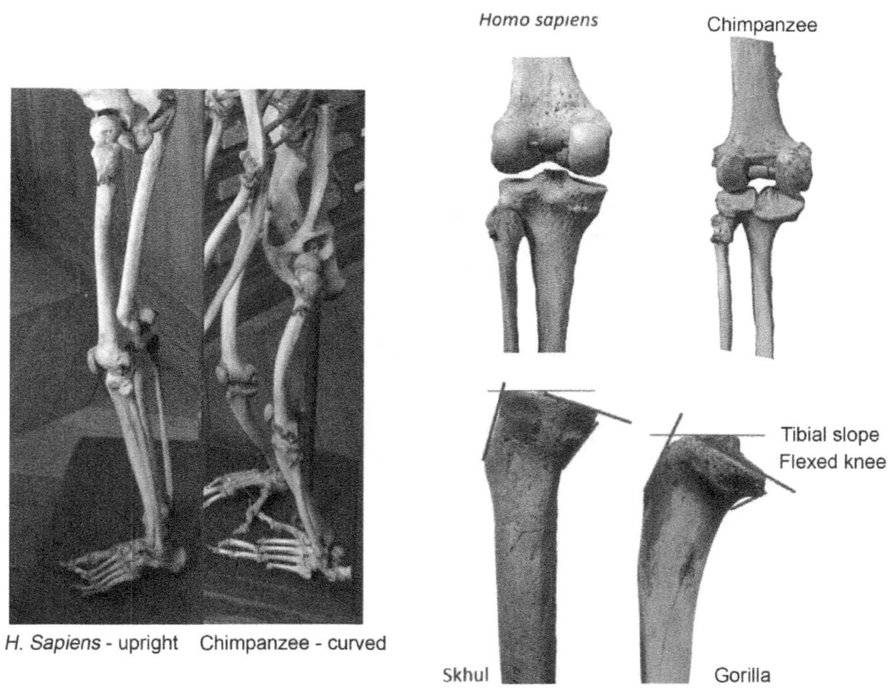

Figure 1.11. *Comparison of femurs between* Homo sapiens *and a chimpanzee. Institut de paléontologie humaine collections (photos and copyright: A. Dambricourt Malassé). For a color version of this figure see www.iste.co.uk/dambricourtmalasse/futureofsapiens.zip*

The psychomotor and occluso-postural development was different, with a higher cerebellar lodge, reciprocal tension membranes extending from the sphenoid and petrous pyramids to the coccyx, and a control of the balance of the head with a lower posterior semicircular canal. Cerebro-spinal tensions forced the coxal bone to adopt an anteversed orientation. The knee joints, compared to *Sapiens* and chimpanzees, show a posteriorly inclined articular surface of the tibial plateau (see Figure 1.11). Not only was the femur curved, but the knees were bent.

The craniosacral posture of the vaulted Chapelle-aux-Saints, as collapsed, is that of a consistent occluso-postural balance that is better understood once we introduce the dynamics of the skull base in extension, and virtually retrace the path of the dorsal chord (see Figure 1.12). Its trajectory shows an almost complete absence of cervical and lumbar lordosis (hypolordosis).

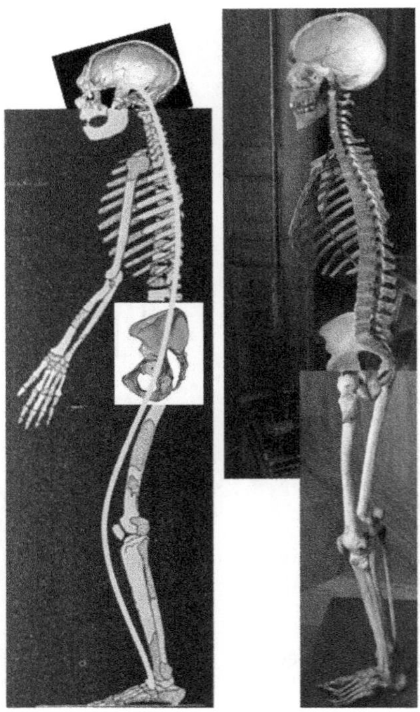

Figure 1.12. *Visualization of basi-craniopodal curvatures of the Neanderthal and* Sapiens *(coxal bone according to Sawyer and Maley (2005), IPH collection, photo and copyright Dambricourt Malassé). For a color version of this figure see www.iste.co.uk/dambricourtmalasse/futureofsapiens.zip*

Our findings have since been corroborated by studies of newborns, of the adult skeleton of Kebara 2 (Israel, 60,000 years), and of the older taxon of *Homo erectus tautavelensis* from Sima de los Huesos (Spain, 430,000 years)[4].

The newborn of Mezmaiskaya (Russian Caucasus, 30,000 years) confirms this: "The face of the Neanderthal newborn is larger and more projected, with the facial characteristics established during fetal growth" (Gunz et al. 2012). The authors observe the classical differences in profile of the neurocranium with *Homo sapiens*:

> The globular braincase of modern humans is distinct from all fossil human species, including our closest extinct relatives, the Neandertals. Such adult shape differences must ultimately be rooted in different developmental patterns, but it is unclear at which point during ontogeny these group characteristics emerge (Gunz et al. 2012).

As long as embryogenesis remains the blind spot of paleoanthropology, this discipline will continue to conclude that "it is not clear"; we have been showing and publishing it for almost 40 years. The moment of emergence begins with embryonic rotation of the post-sphenoid and the associated connective membrane tensions. The sphenoid of this newborn is preserved, which is quite rare, but no study of it has been carried out. The ontogenetic stage was therefore missing in order to understand the intrauterine origin of the differences between the craniofacial morphogeneses of Neanderthal and *Homo sapiens*.

The newborn of Moustier 2 (France, 40,000 years) also confirms this: "Our reconstruction of Moustier 2 shows that most facial differences between modern humans and Neanderthals develop before birth" (Gunz et al. 2011). The face is prognathic, the sphenoid is preserved, but has not been studied.

The deduced posture of the adult skeleton of Kebara 2 (sacrum has been reconstructed) (Sawyer and Maley 2005) is consistent with that of La Chapelle-aux-Saints (Been et al. 2017) and leads to the observation of the inversion of its phylogenetic trajectory relative to that of *Sapiens*:

> When Neandertal lineage hominins are ignored, the results indicate a fairly simple path of evolution from nonhuman hominoid-like to human-like sacral orientation with two stages of the development. Neandertal lineage hominins show a reversal of this trend (Been et al. 2013).

4 In May 2023, during a seminar at the Institut de Paléontologie Humaine in Paris, Juan Luis Arsuaga recognized the historical anteriority of the taxon *tautavelensis* (Lumley 2015) to qualify these hominids by downgrading them as *Homo heidelbergensis*.

> Neandertals, when compared to modern humans, show a smaller degree of lumbar lordosis. Based on a still limited fossil sample, early hominins (australopiths and Homo erectus) had a lumbar lordosis that was similar to but below the mean of modern humans. Here, we hypothesize that from this ancestral degree of lumbar lordosis, the Neandertal lineage decreased their lumbar lordosis and *Homo sapiens* slightly increased theirs (Gómez-Olivencia et al. 2017).

This hypothesis has been verified since our previous conclusions. The occluso-postural dynamics of the Neanderthals has drifted toward extension (Dambricourt Malassé 1992, 1993, 2002; Dambricourt Malassé et al. 2025); this process is also observed in Southeast Asia with *Homo erectus* lineages (e.g. Sangiran 17) (Dambricourt Malassé 2002, p. 394, Figure 7).

The rib cages of children at Mezmaiskaya 1, Le Moustier 2, Dederiyeh 1 (Syria, 70,000–50,000 years, two years old), Roc de Marsal (France, 70,000–60,000 years, between two and four years old) also present prenatal differences:

> The deep and short ribcage of the Neanderthals was already present at birth, as were other skeletal species-specific traits. This morphology possibly represents the plesiomorphic condition shared with *Homo erectus* [...]. This is consistent with the idea that the Neanderthal body plan is likely plesiomorphic in the genus *Homo*, inherited at least from their Middle Pleistocene ancestors from Sima de Los Huesos, if not already from *H. erectus* (García-Martínez et al. 2020).

The large width of the bell-shaped, not barrel-shaped, rib cage is consistent with the width between the two iliac wings of the pelvis. It is interpreted as a possible adaptation of the kidneys and liver during a glacial winter period, to a depletion of the fat- and carbohydrate-rich resources provided by the big game. The body would have compensated for it through high consumption of protein as an energy source, generating an increase in the activity, weight, and probably volume of the liver and the urinary system (Ben-Dor et al. 2016). The large width of the thorax may also correspond, without contradiction, to a developed heart and lungs (see Chapter 2).

1.5.3. *Embryonic dynamics, homeotic genes and craniosacral reciprocal tension membranes in the Neanderthal*

A recent reconstruction of La Chapelle-aux-Saints would contradict its posture being more curved than *Sapiens*:

> La Chapelle-aux-Saints exhibits a pelvic incidence (and hence lumbar lordosis) similar to modern humans, articulation of lumbar and cervical vertebrae indicating pronounced lordosis. Our findings challenge the view of generally small spinal curvatures in Neandertals. Setting aside the developmentally abnormal Kebara 2 vertebral column, La Chapelle-aux-Saints 1 is joined by other Neandertals with sufficient vertebral remains in providing them with a fully upright (and human) axial posture (Haeusler et al. 2019).

However, this reconstruction does not integrate basi-craniofacial architectonics in extension, or knee joints in flexion. It is not compatible with the sphenosacral balance stresses that would be imposed by such a spine. It also lacks consistency because the vertical plane of the TMJ passes behind the lumbar lordosis. Lastly, it is invalidated by several physiological contradictions.

The new axial body plan of *Sapiens* and the rostro-caudal tensions acquired during embryogenesis allow it to erect a degree of verticality and a craniosacral posture that earlier hominins, and thus Neanderthals in particular, could not reach. Indeed, on a spinal level, the establishment of a cervical lordosis is conceivable only at the cost of a conjugation of morphogenetic and physiological mechanisms of the skull base: anterior ascent of the basi-sphenoid and the basi-occipital, hence a greater clivus slope associated with a greater rotation of the spheno-occipital synchondrosis (SOS), a descent of the cerebellar lodge and a mesial drift of the foramen magnum. This new configuration confers in the sphenoid-sacral verticality that adapts to it a gravitational load line passing in front of the C7/T1 vertebrae, behind T12/L1, positioning the sacrum in anteversion and behind the tangent of the femoral axis. In Neanderthals, the characteristics of the skull base are plesiomorphic, but they are also in extension, i.e. with dynamics of the sphenobasilar slope inverse to *Sapiens*, with a foramen magnum in a more dorsal position, which logically confers a cervical vertebral column, i.e. if not kyphotic, then at least straightened (Hadjouis 2021).

In this case, the gravitational load line must pass well in front of the thoracolumbar vertebral column (see Figure 1.12). Weakness or absence of lumbar lordosis will invariably lead to postural stresses: retroversion of the sacrum and semi-flexion of the knee with an exaggerated tibial slope (17° in the Neanderthal of Spy), followed by an internal rotation of the tibial compartments. As an indication, the tibial slope is of the order of 15° on modern Skhul specimens, and 14° on that of Taforalt (see Figure 1.13(a)), while the tibial slope is greatly inclined in the gorilla (26°). The short, curved conformation of the Neanderthal femoral diaphysis (see Figure 1.13(b)) comes under biomechanics and stresses that are axial endo- and exo-craniosacral in flexion.

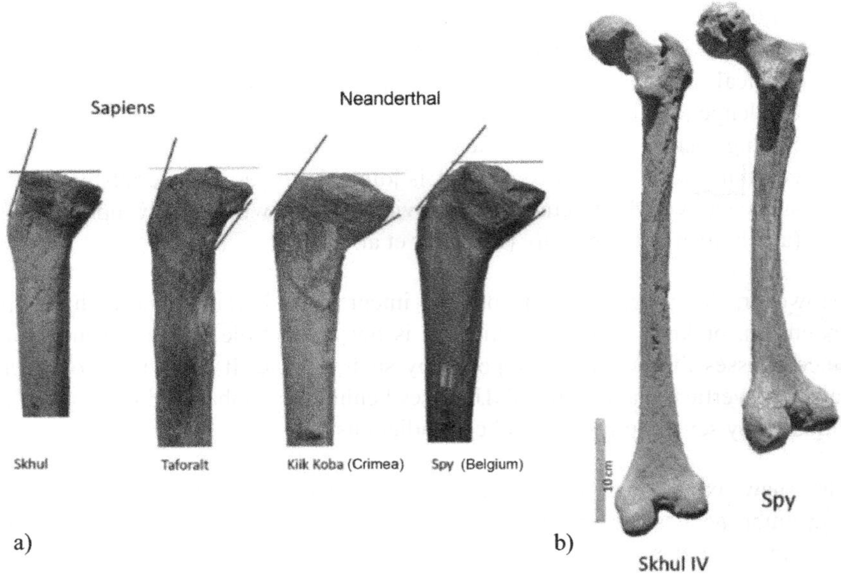

Figure 1.13. *a) Comparison of tibial slope inclinations, between fossil* Homo sapiens *and* Homo neanderthalensis. *b) Comparison in the dorsal view of the femurs of fossil* Homo sapiens *(Skhul IV) and* Homo neanderthalensis *(Spy) (casts of Institut de paléontologie humain collections, photos and copyright: A. Dambricourt Malassé). For a color version of this figure see www.iste.co.uk/dambricourtmalasse/futureofsapiens.zip*

Taking into account these postural models, the attempt at reconstruction by Haeusler et al. (2019) does not seem to us to be reliable, as from the outset, it projects a cervical vertebral column in lordosis without taking into account the extensive cranio-occipital dynamics, while the lower cervical vertebral column is flanked by significant flattening of the anterosuperior edges, accompanied by cervical arthrosis, precisely in the direction of cervical kyphosis. Similarly, reconstructed hyperlordosis gives the sacrum projection an excessively receded anteversion.

The study by Marcellin Boule is meticulous and of rare anatomical and physiological clarity, his reconstruction of the vertebral column and the lower limb being entirely accurate. Criticism of his reconstruction is moreover not supported by

> Regarding the lumbar curvature, nonpathological pelvic and lumbar specimens from the SH site and Neandertal individuals are consistent with the suggestion that the hypolordotic lumbar column in

> Neandertals was already present in their Middle Pleistocene ancestors (Bonmati et al. 2010).

NOTE.– The general external appearance of the Neanderthal is described as that of a stocky being, with a low center of gravity, a powerful neck, and wide oral and nasal cavities. In the internal view, it corresponds to a decrease in the embryonic dynamics of the craniocaudal membranes relative to older species, which results in a general "sagging" of the bony framework up to the position of the posterior semicircular canal. There would have been a decrease in energy in the connective tissues (the actomyosin complex). The regulation of the homeotic genes that control the migration of rhombomere neural crest cells has undergone hereditary alterations. These alterations occurred either during ovocyte gametogenesis (the germline of embryogenesis; they are believed to be genetic) or after ovulation and fertilization until the zygote's implantation in the intrauterine wall (they are believed to be epigenetic). In all cases, the medium is the maternal body, which precedes postnatal functions. Recent studies support this view:

> Skeletal differences between Neandertals and modern humans are largely established by the time of birth. Features that are present on the skeletons of neonatal Neandertals could not have developed in response to mechanical loading associated with adult/older-subadult behaviors (e.g. hunting), so they must either have an evolutionary explanation or result from environmental influences on the fetus (e.g. maternal diet) (Weaver et al. 2016).

Fetal growth may have been impacted by the living conditions of mothers in an ice age, and may have been detrimental to them. Jean Chaline (2003), a multidisciplinary paleontologist, proposed the eclampsia hypothesis to understand the extinction of the Neanderthals:

> Preeclampsia occurs in the second or third trimester of pregnancy. It is a dysfunction of the placenta, which releases into the maternal circulation high concentrations of molecules that become toxic, particularly to the kidneys and the blood-vessel walls. Associated with high blood pressure and the appearance of protein in the urine, this placental dysfunction also affects the baby, who will often experience growth retardation in utero[5].

5 Camilleri (2023). *Grossesse santé contre la pre-éclampsie*. Pregnancy health association for the prevention of preeclampsia. See: https://grossesse-sante.org (or English-language equivalent, https://www.preeclampsia.org, for example).

It is here that the question of viable hybridization between *Sapiens* and Neanderthals arises, while embryogenesis is not taken into consideration. However, it seems unlikely that the genome that codes for the reorganized and more dynamic embryogenesis of *Sapiens* can viably pair with that of Neanderthals, as the regulations of homeotic genes have drifted in the opposite "direction", relative to ancestral embryogenesis (Dambricourt Malassé 2011). The few shared genes may have been transmitted through hybridization between African populations shortly before the division between the Neanderthal ancestral lineage and that of *Sapiens*. This viewpoint is shared by geneticists (Eriksson et al. 2012; Yang et al. 2012):

> Neanderthal and *Homo sapiens* may have retained common traits (identical snip alleles) in their DNA without having interacted genetically. The question is not as clear-cut as the media impact of the works of S. Pääbo might lead one to believe. Moreover, the conclusion of the article is much less categorical than its title and does not exclude the possibility of a significant contribution from the ancient structure of the populations (Jordan 2012).

The doubt as to a viable hybridization is also supported by recent research on the simulation of spatial structures of Neanderthal and *Sapiens* populations (Tournebize and Chikhi 2024).

1.6. Conclusion

The reconstruction of the occluso-postural development of Neanderthals is necessarily placed in the perspective of its phylogenetic ancestry from the first African *Homo* species. Bipedalism was not the triggering factor of the hominization process (verticalization), the size of the brain is not the Rubicon that separates a *Homo* species from other hominins, and the increase in encephalic volume (brain–cerebellum) was not the mechanical cause of neck winding. The verticalization and complexification of the encephalon, which includes the cerebellum and the brain stem – participated in the same active process from fertilization and constitute precisely what fundamental research, concerned with the future of our descendants, should seek to understand:

> This phenomenon is of course observable only in the long term, but it reflects a logic of astonishing stability, an irreversible time that memorizes! The sense of the history of inert, living, thinking matter, in its race toward complexity, is a troubling observation (Coppens 1996).

This "race toward complexity" is an observation for the central nervous system (Lumley and Lledo 2019). The timescale of evolutionary processes refers to the most vital constraints for organisms: the search for balance and the transmission of its mechanisms, and the search for stability and its reproducibility, are the principles of maintaining life. These are the laws of thermodynamics present at all scales of organization, from gametogenesis to fertilization, until the end of life (Queiros-Condé et al. 2022). The system is in equilibrium with itself and its environment, from the placenta to maternal and social behaviors, placed within environmental conditions according to complex processes of self-organization. Ilya Prigogine has shown what the conditions are that allow a system to "break the circle" of reproducibility, allowing it to raise its level of organization (of complexities of interactions), the latter needing to be distanced from its balance state (Prigogine and Stengers 1978).

NOTE.– Irreversibility becomes a constraint if the system again reaches threshold limits of instabilities. It will no longer be able to return to its original stable state, innovation being the condition for its survival within the limits allowed by its organization; otherwise, it is extinction.

The increase in axial complexities and instabilities is the signature of the process of verticalization of the nervous system, perceptible on a decline over 40 million years, from stable horizontal balance to the most unstable vertical balance, ours. It is also the most complex balance to control for the balance control center, the cerebellum, since the latter finds itself in the loop.

Embryogenesis reveals that, ultimately, it is the future position of the cerebellum and the brain stem relative to the axis of gravity, which is at stake for the survival of the new organism, from the first weeks after fertilization, not only with the duration of migrations of the neural crest cells but also with the establishment of the reciprocal tension membranes. Increasing interest in the cerebellum has shown that the connectivity of prefrontal neural networks with the neocerebellum developed from the great apes (Magielse et al. 2023), therefore following the second straightening (20 million years). It has also been shown that over the past 10,000 years, the volume of the cerebellum has become larger in absolute value, while that of the brain has declined (Weaver 2005). Therefore, cerebellization did not stop at the last threshold of *Sapiens* straightening, manifesting between 300,000 and 230,000 years ago, while its role in cognitive functions is recognized to the point of evoking a conceptual revolution:

> The central position of the cerebellum, within which all intrinsically-connected motor, cognitive and affective networks converge, in return offers it the possibility not only of specifically influencing one mental domain or another, but also of coordinating and integrating its functioning closely and dynamically. The cerebellum would play a general modulatory role allowing control, correction, optimization and

> automation of cerebral operations, whose specific functional domain would depend upon the specificity of the cerebello-cortical connections. In thirty years, a conceptual revolution in the approach to the cerebellum has occurred (Habas 2022).
>
> This important finding ran parallel to a change in research focus on the cerebellum – from a selective interest in motor processes to a growing awareness that the human cerebellum supports major cognitive, affective, and social functions (Van Overwalle et al. 2024).

And, as we might have expected, the vestibular system would be involved in the cognitive functions of the cerebellum (Lopez et al. 2023, 2024):

> Even if the vestibular system is far from having delivered all its secrets, science is now able to understand that it is involved in high-level functions such as cognition, the representation of the self in space or the self. It therefore plays an essential role in the well-being and development of each individual (Lopez et al. 2023).

The differentiation of the Neanderthal lineage thus demonstrates at what level of organization of life the processes of morphogenesis are situated after exceeding a threshold of rotation. It is indeed a question of thermodynamic balances and complexity at work since gametogenesis, with risks of energy dissipation. Decreases in tension during winding–unwinding movements of the embryo in formation seem to us a plausible explanation to explore.

Research on the embryogenesis of the different current species of chordae show that the regulation changes concern homeotic genes and therefore the cells of the neural crest, according to the craniocaudal gradient, and that it was therefore possible for reproduction cells to transmit mechanisms for self-organization of the bodily organization plan. In contrast, the genetics of Evo-Devo lack the geological timescale to understand how certain germ lines have been capable, after millions of years of stability, of transmitting, and amplifying, complexifications of physico-chemical interactions capable of reconstructing the bodily organization plan. The increase in the amplitude of rotation of the sphenoid, observed during a change in plan, can be interpreted as the effect of prolongations of the duration of NCC migrations. These heterochronies are found throughout fetal and postnatal development, to the benefit of the central nervous system, longer psychomotor development and later puberty.

The interpretation of the phylogenesis of *Sapiens* can therefore no longer afford not to consider embryogenesis, intrauterine life, living conditions of mothers and advances in knowledge in the sciences of self-organization (Dambricourt Malassé 2022c, 2022d). This curve appears exponential and irreversible; the cerebellum has supplanted the neocortex for 10,000 years. In such a context, little is known in DFO, the high percentage of agenesis of dental ends-of-series, i.e. 27% of lateral upper incisors, 25% of third molars, and 15–32% of second lower premolars[6], should capture our attention. It no longer seems conceivable to reason as to the postural future of our species, which has reached the "wall of internal verticality", without placing the emergence of our embryogenesis on the curve of these increasingly complex interaction processes, and without questioning the circumstances that reactivated the properties of our gametes upstream of neurogenesis complexification, with its effects that lower the cerebellum, reduce the length of the dental arches and modify the cerebro-cerebellar neural networks whose cognitive and emotional functions are now recognized.

1.7. References

Aoudia-Chouakri, L. (2003). Analyse architecturale par téléradiographie des crânes de la grotte de Taforalt (Pléistocène supérieur, Maroc). Comparaison avec les hommes d'Afalou Bou Rhummel (Paléolithique supérieur d'Algérie). Muséum national d'histoire naturelle, Paris.

Autet, B.M. (1985). Examen ostéopathique prenant en compte l'activité tonique posturale. *Mémoire de la Sereto*, Montpellier.

Balasubramaniam, L., Doostmohammadi, A., Saw, T.B., Narayan, G.H., Narayana, S., Mueller, R., Dang, T., Thomas, M., Gupta, S., Sonam, S. et al. (2021). Investigating the nature of active forces in tissues reveals how contractile cells can form extensile monolayers. *Nature Materials*, 20, 1156–1166. doi: 10.1038/s41563-021-00919-2.

Been, E., Pessah, H., Peleg, S., Kramer, P. (2013). Sacral orientation in hominin evolution. *Advances in Anthropology*, 3, 133–141. doi: 10.4236/aa.2013.33018.

Been, E., Gómez-Olivencia, A., Kramer, P.A., Barash, A. (2017). 3D reconstruction of spinal posture of the Kebara 2 Neanderthal. In *Human Paleontology and Prehistory*, Marom, A. and Hovers, E. (eds). Springer, Cham, 239–251. doi: 10.1007/978-3-319-46646-0_18.

Ben-Dor, M., Gopher, A., Barkai, R. (2016). Neandertals' large lower thorax may represent adaptation to high protein diet. *American Journal of Physical Anthropology*. doi: 10.1002/ajpa.22981.

Bolk, L. (1926). Le problème de l'anthropogenèse. *Comptes rendus de l'Association des anatomistes de langue française*, 1, 80–92.

Bonmati, A., Gómez-Olivencia, A., Arsuaga, J.L., Carretero, J.M., Gracia, A., Martínez, I., Lorenzo, C., Bérmudez de Castro, J.M., Carbonell, E. (2010). Middle Pleistocene lower back and pelvis from an aged human individual from the Sima de los Huesos site, Spain. *PNAS*, 107(43), 8386–1839. doi: 10.1073/pnas.1012131107.

6 French National Diagnostic and Treatment Protocol (2021).

Boule, M. (1911). L'homme fossile de la Chapelle-aux Saints. *Annales de paléontologie*, VI.

Chaline, J. (2003). Increased cranial capacity in hominid evolution and preeclampsia. *Journal of Reproductive Immunology*, 131–144.

Chaline, J. and Marchand, D. (2010). *Le singe, l'embryon et l'homme. Une nouvelle clé de lecture de l'histoire de l'homme*. Ellipse, Paris.

Chaline, J., Dambricourt Malassé, A., David, B., Magniez-Jannin, F., Marchand, D., Courant, F., Millet, J.J. (1998). Quantification de l'évolution morphologique du crâne des hominidés et hétérochronies. *Comptes Rendus de l'Académie des Sciences, Séries IIA, Sciences de la Terre et des planètes/Earth and Planetary Science*, 326(4), 291–298. doi: 10.1016/S1251-8050(97)86820-6.

Cho, K.H. (2005). Asymétries cranio-faciales et dysmorphoses. Analyse architecturale cranio-faciale des populations médiévales de La Queue-En-Brie (Val-de-Marne, France). PhD Thesis, Muséum national d'Histoire naturelle, Paris.

Cho, K.H. and Hadjouis, D. (2005). Les asymétries cranio-faciales de la nécropole médiévale de la Queue-En-Brie (Ile-de-France, Val-de-Marne). *Biométrie humaine et anthropologie*, 23(1/2), 95–104.

Clauzade, M.A. and Darraillans, B. (1992). *L'homme, le crâne, les dents*. SEOO Editions, Perpignan.

Clauzade, M.A., Darraillans, B., Knellesen, C., Montet, G. (1989). *Concept ostéopathique de l'occlusion*. SEOO Editions, Perpignan.

Clauzade, M., Clauzade, N., Ouhioun, J.L. (2016). *Le capteur mandibulo-spinal*. Itimedias Éditions, Saint Herblain.

Conde Vázquez, O. (2019). Étude des influences mandibulaires sur le test des rotateurs de hanche chez l'enfant. *Neurophysiologie clinique*, 49(3), 254. doi: 10.1016/j.neucli.2019.01.015.

Conde Vázquez, O. and Suárez-Quintanilla, J. (2020). Anatomical and physiological considerations in the "Hip Rotators Test" related to the stomatognathic system. A brief commentary. *International Journal of Morphology*, 38(2), 363–366.

Coppens, Y. (1996). *Annuaire du Collège de France 1995–1996. Résumés des cours et travaux, 96e année*. Collège de France, Paris.

Dambricourt Malassé, A. (1987). Ontogenèses, paléontogenèses et phylogenèse du corps mandibulaire catarhinien. Nouvelle interprétation de la mécanique humanisante (théorie de la foetalisation, Bolk 1926). PhD Thesis, Muséum national d'Histoire naturelle, Paris.

Dambricourt Malassé, A. (1988). Hominisation et fœtalisation. *Comptes Rendus de l'Académie des Sciences, Séries II, Paléontologie humaine/Human Paleontology*, 307, 199–204.

Dambricourt Malassé, A. (1992). Droit de réponse. *Revue d'orthopédie dento-faciale*, 26(1), 71–79.

Dambricourt Malassé, A. (1993). Continuity and discontinuity during hominization. *Quaternary International*, 19, 85–98.

Dambricourt Malassé, A. (1996). Nouveau regard sur l'origine de l'Homme. *La Recherche*, 286, 46–54.

Dambricourt Malassé, A. (1999). De la morphogenèse cranio-faciale embryonnaire à l'architecture adulte. Problèmes et méthodes de la reconstitution des ontogenèses fossiles. *Mémoires de l'Académie des sciences, Arts et Belles-Lettres de Caen*, 37, 7–81.

Dambricourt Malassé, A. (2002). Nouvelles approches de l'évolution crânienne des Homo erectus de Java. In *Origine des peuplements et chronologie des cultures paléolithiques dans le Sud-Est asiatique*, Sémah, F., Falguères, C., Grimaud Hervé, D., Sémah, A.M. (eds). Artcom, Paris, 377–396.

Dambricourt Malassé, A. (2005). Un nouvel angle de la base pour une meilleure compréhension de la morphogenèse basi-crânienne des hominidés et des modalités évolutives. *Biométrie et anthropologie de la tête, de la face et du cou. Biométrie humaine et anthropologie*, 23(1/2), 53–62.

Dambricourt Malassé, A. (2010). The relationships between occlusion and posture in the hominid lineage, implications for the transition between Mesolithic and Neolithic populations. Original synthetic article. *International Journal of Modern Anthropology*, 3, 14–63.

Dambricourt Malassé, A. (2011). Les équilibres bipèdes permanents, origine embryonnaire, morphogenèse, équilibre occluso-postural, conséquences pour l'évolution psychomotrice et comportementale des hominidés. HDR Thesis, Université de technologie de Compiègne, Compiègne.

Dambricourt Malassé, A. (2017). The last threshold. Inference. *International Review of Science*, 3(1).

Dambricourt Malassé, A. (ed.) (2022a). Sapiens and cognition: The optimal vertical nervous system. The last threshold of self-organized and self-memorizing increasing complexity from gametes to embryo. In *Self-Organization as a New Paradigm in Evolutionary Biology, from Theory to Applied Cases in the Tree of Life*. Springer, Cham, 307–358.

Dambricourt Malassé, A. (2022b). *Embryogeny and Phylogeny of the Human Posture 2. A New Glance at the Future of our Species*. ISTE Ltd, London and Wiley, Hoboken, NJ.

Dambricourt Malassé, A. (ed.) (2022c). Introduction. Understanding the origins and evolution of living organisms – The necessity of convergence between old and new paradigms. In *Self-Organization as a New Paradigm in Evolutionary Biology, from Theory to Applied Cases in the Tree of Life*. Springer, Cham, 1–7.

Dambricourt Malassé, A. (2024). La cérébellisation : perspective d'avenir pour Sapiens aux limites de la verticalisation ? In *25ᵉ Journées lilloises de neuropédiatrie*. Centre de référence des maladies et malformations congénitales du cervelet du CHU de Lille, Lille.

Dambricourt Malassé, A. and Deshayes, M.J. (1989). A reconsideration of the hominid evolution by the ontogenetic approach and the nonlinear dynamical systems. Implications and model for evolution of language. In *5th Annual Meeting of Language Origins Society*. Flawn Academic Center, University of Texas, Austin.

Dambricourt Malassé, A. and Deshayes, M.J. (1992). Modeling of the craniofacial architecture during ontogenesis and phylogenesis. In *The Head-Neck Sensory Motor System*, Berthoz, A., Graf, P.P., Vidal, W. (eds). Oxford University Press, New York/ Oxford, 37–46.

Dambricourt Malassé, A. and Deshayes, M.J. (1995). Données ontogéniques et fonctionnelles des capacités langagières néandertaliennes. Implications pour les interprétations paléo-éthologiques. In *Nature et culture. Relations entre la biologie et l'histoire au cours de l'évolution ancienne de l'Homme*, Otte, M. (ed.). ERAUL, Liège, 951–1003.

Dambricourt Malassé, A. and Lallouet, F. (2009). Étude au scanner des canaux semi-circulaires d'un crâne sec d'anencéphale dépourvu de flexion sphéno-occipitale ; implications pour l'interprétation paléo-embryologique des hominidés fossiles. *Avancées en biométrie humaine. Biométrie humaine et anthropologie*, 27, 1–2.

Dambricourt Malassé, A., Martin, J.P., de Kerviler, E. (2000a). Approche biodynamique des angulations de la base du crâne de l'embryogenèse à la morphogenèse postnatale. In *L'identité humaine en question. Nouvelles problématiques et nouvelles technologies en paléontologie humaine et en paléoanthropologie biologique*, Andrieux, P., Hadjouis, D., Dambricourt Malassé, A. (eds). Artcom', Paris, 24–43.

Dambricourt Malassé, A., Martin, J.P., de Kerviler, E. (2000b). Neural tube, spheno-occipital flexion and semi-circular canals in fossil and modern hominids. *Acta Anthropologica Sinica*, 19, 82–89.

Dambricourt Malassé, A., Autet, B., Joffroy S., Hadjouis, D. (eds) (2025a). La verticalité, fil directeur de nos origines et de notre devenir. In *L'avenir de Sapiens et la verticalité*. ISTE Editions, London.

Dambricourt Malassé, A., Ho Ba Tho, M.-C., Dao, T.T., Lallouet, F. (2025). Reconstruction of the internal cranial base of an adolescent Homo erectus (Sinanthropus III). An enlightening on the hominids neural straightening and its cognitive consequences. *L'Anthropologie*, 3. doi: 10.1016/j.anthro.2024.103347.

Dance, A. (2021). The secret forces that squeeze and pull life into shape. *Nature* [Online]. Available at: nature.com/articles/d41586-021-00018-x.

Deshayes, M.J. (1986). *Croissance cranio-faciale et orthodontie*. Masson, Paris.

Deshayes, M.J. (1991). Reconsidération de la croissance cranio-faciale au cours du développement et de l'évolution. Conséquences en orthopédie dento-maxillo-faciale. *La Revue d'Orthopédie Dento-Faciale*, 25(3), 353–365. doi: 10.1051/odf/1991018.

Deshayes, M.J. (2006). Croissance crânienne et cinétiques osseuses interférant avec la morphogenèse faciale. Bases conceptuelles de réussite des traitements orthopédiques avant l'âge de 6 ans. *L'Orthodontie Française*, 77, 63–78. doi: 10.1051/orthodfr/200677063.

Deshayes, M.J. and Dambricourt Malassé, A. (1991). Analyse des différents types architecturaux cranio-faciaux par l'approche ontogénique de l'hominisation. *Revue de stomatologie et de chirurgie maxillo-faciale*, 4, 249–258.

Deshayes, M.J. and Deshayes, J. (2013). Étude des relations entre la morphologie cranio-faciale et l'occlusion en denture temporaire : remodelages occipitaux et singularités basicrâniennes déterminantes pour l'occlusion. *L'Orthodontie Française*, 84(1), 97–111. doi: 10.1051/orthodfr/2013036.

Deshayes, M.J., Dambricourt Malassé, A., Deshayes, P. (1990). Utilisation de la téléradiographie dans la compréhension de la mécanique humanisante. In *L'image et la science, 115*[e] *Congrès national des sociétés savantes*. Editions du Comité des travaux historiques et scientifiques (CTHS), Paris, 149–157.

Duboule, D. (2020). Évolution des génomes et développement. *L'annuaire du Collège de France. Cours et travaux 2017–2018*, 118, 569–573.

Eriksson, A. and Manica, A. (2012). Effect of ancient population structure on the degree of polymorphism shared between modern human populations and ancient hominins. *PNAS*, 109, 13956–13960.

Fekete, D.M., Homburger, S.A., Waring, M.T., Riedl, A.E., Garcia, L.F. (1997). Involvement of programmed cell death in morphogenesis of the vertebrate inner ear. *Development*, 124, 2451–2461. doi: 10.1242/dev.124.12.2451.

Ferré, J.-C. and Salagnac, J.-M. (1996). La mobilité des os du crâne chez l'adulte et l'adolescent : une théorie scientifiquement discutable. *Annales de Kinésithérapie*, 23(7), 310–317.

Fleury, V. (2024). Electrical stimulation of chicken embryo development supports the inside story scenario of human development and evolution. *Scientific Reports*, 14, 7250. doi: 10.1038/s41598-024-56686-y.

Fleury, V., Nicolas, R.C., Furfaro, F., Duband, J.-L. (2015). Buckling along boundaries of elastic contrast as a mechanism for early vertebrate morphogenesis. *The European Physical Journal E*, 38, 6. doi: 10.1140/epje/i2015-15006-7.

Freyer, L., Aggarwal, V., Morrow, B.E. (2011). Dual embryonic origin of the mammalian otic vesicle forming the inner ear. *Development*, 138(24), 5403–5414. doi: 10.1242/dev.069849.

Gagey, P.-M. and Weber, B. (2005). *Posturologie. Régulation et dérèglements de la station debout*. Masson, Paris.

García-Martínez, D., Bastir, M., Gómez-Olivencia, A., Maureille, B., Golovanova, L., Doronichev, V., Akazawa, T., Kondo, O., Ishida, H., Gascho, D. et al. (2020). Early development of the Neanderthal ribcage reveals a different body shape at birth compared to modern humans. *Science Advances*, 6, eabb4377.

Gómez-Olivencia, A., Arlegi, M., Barash, A., Stock, J.T., Been, E. (2017). The Neandertal vertebral column 2: The lumbar spine. *Journal of Human Evolution*, 106, 84–101.

Gudin, R.G. and Ferré, J.C. (1968). *Les rétrognathies à caractère fonctionnel. Encyclopédie médico-chirugicale*. Stomatologie, Paris.

Gudin, R.G. and Godard, B. (1964). *Bases physiologiques et fonctionnelles de la thérapeutique mécanique et orthopédie dento-faciale. Encyclopédie médico-chirugicale.* Stomatologie, Paris.

Gudin, R.G. and Godard, B. (1989). Les troubles dyspnéiques des sphères oro-faciale et pharyngée. Leur influence sur la morphogenèse cranio-faciale et leur rapport avec la posture mandibulaire en période de croissance. *Revue d'Orthopédie Dento-Faciale*, 23, 407–425 [Online]. Available at: https://issuu.com/clubscientifique/docs/vol23_nmr4_1989.

Guimberteau, J.C. (2005). La mécanique du glissement des structures sous cutanées chez l'homme. Mise en évidence d'une unité fonctionnelle : la microvacuole. *Les e-mémoires de l'Académie Nationale de Chirurgie*, 4, 35–42.

Guimberteau, J.C., Schleip, R., Findley, T. (2012). The subcutaneous and epitendinous tissue behavior of the multimicrovacuolar sliding system. In *Fascia: The Tensional Network of the Body*, Stecco, C., Driscoll, M., Huijing, P., Schleip, R. (eds). Elsevier, London, 143–146.

Gunz, P., Neubauer, S., Maureille, B., Hublin, J.J. (2011). Virtual reconstruction of the Le Moustier 2 newborn skull. Implications for Neandertal ontogeny/Sur la reconstitution virtuelle du crâne du nouveau-né néandertalien Le Moustier 2. Implications ontogéniques. *Paleo*, 22, 155–172. doi: 10.4000/paleo.2107.

Gunz, P., Neubauer, S., Golovanova, L., Doronichev, V., Maureille, B., Hublin, J.J. (2012). A uniquely modern human pattern of endocranial development. Insights from a new cranial reconstruction of the Neandertal newborn from Mezmaiskaya. *Journal of Human Evolution*, 62(2), 300–313. doi: 10.1016/j.jhevol.2011.11.013.

Habas, C. (2022). Cervelet et cognition. Cerebellum and cognition. *Bulletin de l'Académie Nationale de Médecine*, 206(1), 38–49.

Hadjouis, D. (1999). *Les populations médiévales du Val-de-Marne. Dysharmonies cranio-faciales, maladies bucco-dentaires et anomalies du développement dentaire au cours du Moyen Âge.* Artcom', Paris.

Hadjouis, D. (2000a). Équilibre ou déséquilibre, normoclusion ou malocclusion. Actualités odonto-stomatologiques. *L'Encyclopédie du praticien*, 213, 35–45.

Hadjouis, D. (2000b). Étiopathogénies des dysharmonies cranio-faciales de populations médiévales du Val-de-Marne. In *L'identité humaine en question. Nouvelles problématiques et nouvelles technologies en paléontologie humaine et en paléoanthropologie biologique*, Andrieux, P., Hadjouis, D., Dambricourt Malassé, A. (eds). Artcom', Paris, 170–186.

Hadjouis, D. (2002). Les hommes du paléolithique supérieur d'Afalou-Bou-Rhummel (Bedjaia, Algérie). Interprétation nouvelle des cinétiques cranio-faciales et des effets de l'avulsion dentaire. Malformations crâniennes, troubles de la croissance, anomalies et maladies alvéolo-dentaires. *L'Anthropologie*, 106, 337–375.

Hadjouis, D. (2003). Hominidés et grands mammifères dans leur contexte paléoenvironnemental au cours du Quaternaire maghrébin. HDR Thesis, Université de Perpignan, Perpignan.

Hadjouis, D. (2011). Les hommes de Mechta-Afalou d'Algérie. Architecture crânio-faciale, occlusion biodynamique, paléopathologie. L'exemple des hommes paléolithiques d'Afalou-Bou-Rhummel. Thesis, CNRPAH, Alger.

Hadjouis, D. (2015). Peuplements de l'Afrique du Nord. *L'encyclopédie berbère*, 37, 6034–6052.

Hadjouis, D. (2017). Forte fréquence de malformations architecturales cranio-faciales et de classe III dans une population médiévale du sud-est parisien (La Queue-en-Brie, Val-de-Marne). *L'Orthodontie bioprogressive*, 8–11.

Hadjouis, D. (2018). *Atlas des maladies et traumatismes du monde médiéval et moderne, Ve siècle-XVIIe siècle*. ISTE Editions, London.

Hadjouis, D. (2021). *The Skull of Quadruped and Bipedal Vertebrates: Variations, Abnormalities and Joint Pathologies*. ISTE Ltd, London and Wiley, Hoboken, NJ.

Hadjouis, D., Vignal, J.N., Medig, M., Sahnouni, M., Derradji, A. (2000a). Biodynamique cranio-faciale, paléopathologie et reconstitution faciale 2D d'un crâne ibéromaurusien (grotte de Taza 1, Jijel, Algérie). In *L'identité humaine en question. Nouvelles problématiques et nouvelles technologies en paléontologie humaine et en paléoanthropologie biologique*, Andrieux, P., Hadjouis, D., Dambricourt Malassé, A. (eds). Artcom', Paris, 205–220.

Hadjouis, D., Vignal, J.N., Michaut, J.-F., Richebé, J., Schuliar, Y. (2000b). Analyse architecturale, paléopathologie et reconstitution faciale 2D et 3D (sculpture) d'un crâne mechtoïde du Néolithique de l'Ouest algérien (Oued Guettara, Algérie). In *L'identité humaine en question. Nouvelles problématiques et nouvelles technologies en paléontologie humaine et en paléoanthropologie biologique*, Andrieux, P., Hadjouis, D., Dambricourt Malassé, A. (eds). Artcom', Paris, 221–236.

Haeusler, M., Trinkaus, E., Fornai, C., Müller, J., Bonneau, N., Boeni, T., Frater, N. (2019). Morphology, pathology and the vertebral posture of the La Chapelle-aux-Saints Neandertal. *Proceedings of the National Academy of Sciences of the USA*, 116(11), 4923–4927. doi: 10.1073/pnas.1820745116.

Heim, J.L. (1989). La nouvelle reconstitution du crâne néandertalien de la Chapelle-aux-Saints. Méthodes et résultats. *Bulletins et Mémoires de la Société d'Anthropologie de Paris*, 1(1/2), 95–117.

Heim, J.L., Hadjouis, D., Aumassip, G. (2011). L'enfant néolithique de Tin Hanakaten (Homo n°5) (Tassili des Ajjers, Algérie). Examen anthropologique et paléopathologique. *CNRPAH*, 11(II), 27–40.

Hodgson, N. (2006). The role of the spinal dura mater in spinal subluxation. *Semantic Scholar* [Online]. Available at: https://api.semanticscholar.org/CorpusID:29711910.

Ioniță, C., Petre, A.E., Cononov, R.S., Covaleov, A., Mitoiu, B.I., Nica, AS. (2023). Methods of postural analysis in connection with the stomatognathic system. A systematic review. *Journal of Medicine and Life*, 16(4), 507–514. doi: 10.25122/jml-2022-0327.

Izelfanane, H. (2011). Insertions de la dure-mère sur le rachis : une anatomie redécouverte. *La Revue de l'Ostéopathie*, 4(1), 5–15.

Jin, S.W., Sim, K.B., Kim, S.D. (2016). Development and growth of the normal cranial vault: An embryologic review. *Journal of Korean Neurosurgical Society*, 59, 192–196.

Jordan, B. (2012). Neandertal and *Homo sapiens*: To meet, or not to meet? *Med Sci (Paris)*, 28(12), 1129–1132. doi: 10.1051/medsci/20122812025.

Kapandji, A.I. (2012). Le système conjonctif, grand unificateur de l'organisme. *Annales de chirurgie plastique esthétique*, 57(5), 507–514. doi: 10.1016/j.anplas.2012.07.007.

Langman, J. (1984). *Embryologie médicale*. Masson, Paris.

Levi, G. (1900). Beitrag zum studium der entwickelung des knorpeligen primordialcraniums des menschen. *Archiv für mikroskopische Anatomie*, 55, 341–414.

Lopez, C. and Cullen, K.E. (2024). Electrical stimulation of the peripheral and central vestibular system. *Current Opinion in Neurology*, 37(1), 40–51. doi: 10.1097/WCO.0000000000001228.

Lopez, C., Dary, Z., Elziere, M., Perfumo, S. (2023). Le système vestibulaire, un sixième sens méconnu. Video, CNRS [Online]. Available at: https://images.cnrs.fr/video/7936#.

de Lumley, M.A. (2015). L'homme de Tautavel. Un Homo erectus européen évolué. Homo erectus tautavelensis. *L'Anthropologie*, 119(3), 303–348.

de Lumley, H. and Lleido, P.M. (eds) (2019). *Évolution du cerveau et des capacités cognitives des hominidés fossiles*. CNRS Éditions, Paris.

Magielse, N., Heuer, K., Toro, R., Schutter, D.J.L.G., Valk, S.L. (2023). A comparative perspective on the cerebello-cerebral system and its link to cognition. *Cerebellum*, 22(6), 1293–1307. doi: 10.1007/s12311-022-01495-0.

Méndez-Maldonado, K., Vega-López, G.A., Aybar, M.J., Velasco, I. (2020). Neurogenesis from neural crest cells: molecular mechanisms in the formation of cranial nerves and ganglia. *Frontiers in Cell and Developmental Biology*, 8, 635. doi: 10.3389/fcell.2020.00635.

Minoux, M. and Rijli, F.M. (2010). Molecular mechanisms of cranial neural crest cell migration and patterning in craniofacial development. *Development*, 137(16), 2605–2621. doi: 10.1242/dev.040048.

Morriss-Kay, G. and Tuckett, F. (1991). Early events in craniofacial morphogenesis. *Journal of Craniofacial Genetics and Developmental Biology*, 11, 181–191.

Nishikori, T., Hatta, T., Kawauchi, H., Otani, H. (1999). Apoptosis during inner ear development in human and mouse embryos: An analysis by computer-assisted three-dimensional reconstruction. *Anatomy and Embryology*, 200, 19–26. doi: 10.1007/s004290050255.

Paidas, C., Hutchins, G., Morreale, R. (2001). Human embryo growth movement. Morphing video, Johns Hopkins University, Baltimore.

Parker, H.J., Pushel, I., Krumlauf, R. (2018). Coupling the roles of Hox genes to regulatory networks patterning cranial neural crest. *Developmental Biology*, 444(1), S67–S78. doi: 10.1016/j.ydbio.2018.03.016.

Petit-Maire, N. and Charon, M. (1979). Les nécropoles de Tintan et Chami, contexte biologique, 1. Les restes humains. In *Le Sahara atlantique à l'Holocène, peuplement et écologie*, Petit-Maire, N. (ed.). CRAPE/SNED, Alger, 151–227.

Petit-Maire, N. and Fleichier, J.P. (1979). La Sebkha Mahariat. In *Le Sahara atlantique à l'Holocène, peuplement et écologie*, Petit-Maire, N. (ed.). CRAPE/SNED, Alger, 71–90.

Petit-Maire, N., Charon, M., Orthlieb, L. (1979). La dune d'Izriten. In *Le Sahara atlantique à l'Holocène, peuplement et écologie*, Petit-Maire, N. (ed.). CRAPE/SNED, Alger, 17–34.

Prigogine, I. and Stengers, I. (1978). *La nouvelle alliance. Métamorphose de la science.* Folio, Paris.

Queiros-Condé, D., Chaline, J., Brissaud, I. (2022). Quantum fractal thermodynamics to describe the log-periodicity law in species evolution and human organizations. In *Self-Organization as a New Paradigm in Evolutionary Biology*, Dambricourt Malassé, A. (ed.). Springer, Cham, 291–305.

Sawyer, G.J. and Maley, B. (2005). Neanderthal reconstructed. *The Anatomical Record B, New Anatomy*, 283(1), 23–31. doi: 10.1002/ar.b.20057.

Schneider, R.A. (2018). Neural crest and the origin of species-specific pattern. *Genesis*, 56(6/7), e23219. doi: 10.1002/dvg.23219.

Shao, Y., Zhou, L., Li, F. et al. (2023). Phylogenomic analyses provide insights into primate evolution. *Science*, 380(6648), 913–924. doi: 10.1126/science.abn6919.

Smith, T.D., Reynolds, R.L., Mano, N., Wood, B.J., Oladipupo, L., Hughes, G.K., Corbin, H.M., Taylor, J., Ufelle, A., Burrows, A.M. et al. (2021). Cranial synchondroses of primates at birth. *The Anatomical Record*, 4(5), 1020–1053. doi: 10.1002/ar.24521.

Spoor, F. and Zonneweld, F. (1997). The bony labyrinth of Sts 5 ("Mrs Ples"). *South African Journal of Science*, 93, 180.

Spoor, F., Hublin, J.J., Braun, M., Zonneveld, F. (2003). The bony labyrinth of Neanderthals. *Journal of Human Evolution*, 44, 141–165.

Theveneau, E. and Mayor, R. (2012). Neural crest delamination and migration: From epithelium-to-mesenchyme transition to collective cell migration. *Developmental Biology*, 366(1), 34–54.

Utsunomiya, N., Katsube, M., Yamaguchi, Y., Yoneyama, A., Morimoto, N., Yamada, S. (2022). The first 3D analysis of the sphenoid morphogenesis during the human embryonic period. *Scientific Reports*, 28(12), 5259. doi: 10.1038/s41598-022-08972-w.

Van Overwalle, F., Ma, Q., Haihambo, N., Bylemans, T., Catoira, B., Firouzi, M., Li, M., Pu, M., Heleven, E., Baeken, C. et al. (2024). A functional atlas of the cerebellum based on NeuroSynth task coordinates. *Cerebellum*, 23(3), 993–1012. doi: 10.1007/s12311-023-01596-4.

Villedieu, A., Alpar, L., Gaugué, I., Joudat, A., Graner, F., Bosveld, F., Bellaïche, Y. (2023). Homeotic compartment curvature and tension control spatiotemporal folding dynamics. *Nature Communication*, 14(594). doi: 10.1038/s41467-023-36305-6.

Wang, W., Chan, E.K., Baron, S., Van de Water, T., Lufkin, T. (2001). Hmx2 homeobox gene control of murine vestibular morphogenesis. *Development*, 128(24), 5017–5029. doi: 10.1242/dev.128.24.5017.

Weaver, A.H. (2005). Reciprocal evolution of the cerebellum and neocortex in fossil humans. *Proceedings of the National Academy of Sciences of the USA*, 102(10), 3576–3580. doi: 10.1073/pnas.0500692102.

Weaver, T.D., Coqueugniot, H., Golovanova, L.V., Doronichev, V.B., Maureille, B., Hublin, J.J. (2016). Neonatal postcrania from Mezmaiskaya, Russia, and Le Moustier, France, and the development of Neandertal body form. *Proceedings of the National Academy of Sciences of the USA*, 113, 6472–6477.

Yang, M.A., Malaspinas, A.S., Durand, E.Y., Slatkin, M. (2012). Ancient structure in Africa unlikely to explain neanderthal and non-african genetic similarity. *Molecular Biology and Evolution*, 29, 987–2989.

Zheng, N., Yuan, X.Y., Chi, Y.Y. et al. (2017). The universal existence of myodural bridge in mammals: An indication of a necessary function. *Scientific Reports*, 7(8248). doi: 10.1038/s41598-017-06863-z.

Zheng, N., Chung, B.S., Li, Y.L., Liu T.Y., Zhang, L.X., Ge, Y.Y., Wang, N.X., Zhang, Z.H., Cai, L., Chi, Y.Y. et al. (2020). The myodural bridge complex defined as a new functional structure. *Surgical and Radiologic Anatomy*, 42, 143–153. doi: 10.1007/s00276-019-02340-6.

2

Middle Meningeal Vascularization in Neanderthals, and Their Cardiovascular System

2.1. Introduction

The relationship between, on the one hand, the verticalization of the central nervous system that begins at the center of the endocranial base, and, on the other hand, the physiology of the cardiovascular system, the degree of axial verticality and psychomotor development, are little explored in extinct hominins. Yet, this knowledge is necessary in order to understand the evolutionary processes at the origin of lithic *chaînes opératoires*, and therefore, the cognitive and neurosensory capacities of their inventors. The oldest lithic productions have been known since at least 3.3 Ma in Kenya (Harmand et al. 2015) and are attributable to a non-human hominin; the oldest that can be attributed to a very ancient species of the genus *Homo* (Dambricourt Malassé and Cauche 2022) have been known since 2.95 Ma in Asia (Chapon Sao et al. 2024; Dambricourt Malassé et al. 2025c). Understanding the processes of their complexifications occurring in certain circumstances is a multidisciplinary research theme considerably enriched by fossil collections, lithic artifacts and computed tomography technologies applied to the study of the brain (Prat 2024).

The evolution of the cardiovascular system from great apes up to our "modern" anatomy is still little studied. Recent works by Drane et al. (2019) conducted as part of the International Primate Heart Project, have provided leading observations for human paleontology, in discovering that the heart structure of great apes and humans is not uniform. Their research was motivated by the conservation of endangered great

---

Chapter written by Sandra JOFFROY, Anne DAMBRICOURT MALASSÉ and Djillali HADJOUIS.

apes (chimpanzees, bonobos, gorillas and orangutans) and by the high incidence of heart disease in zoos. This research has made it possible to provide veterinary care consistent with their physiology, which differs from that of modern humans. The major differences in structure, size and function are thought to be related to the physical and physiological consequences of verticalization and of the size of the encephalon. Verticalization, contrary to the still-prevalent Lamarckian paradigm, is not locomotive in origin (see the Introduction to the current book). This paradigm does not make it possible to understand, or even to conceive, that the genetic and/or epigenetic, or indeed energetic, evolution at the origin of the morphogenesis and growth of Neanderthals departed from the hominization processes leading to *Homo sapiens* Linnaeus in 1758 or to the anatomy of "modern" humanity at the angular limits of internal verticality. Indeed, if locomotion had been the physical driving force of axial straightening, the embryonic chordal axis of Neanderthals would have to be as verticalized as that of the oldest species of the genus *Homo*. Yet, it is precisely the opposite. The question is therefore not whether the Neanderthals were bipedal – their bipedalism has never been questioned – but by what processes the verticality of their ancestral lineage was diminished, impacting their entire physiology? We therefore propose an attempt to reconstruct the cardiovascular system of Neanderthals and its development, placing our observations relative to this loss of axial verticality in light of recent studies, hypotheses and conclusions. We will focus on the cerebral skull, the rib cage and the craniosacral posture.

2.2. Bone anatomies of the Neanderthal compared to modern anatomy and the extinct taxa of the genus *Homo*: the skull

2.2.1. *The venous network of the meninges*

The formation of the encephalic vascular system of fossil hominins was particularly studied by Roger Saban (1920–2011), professor at the Muséum national d'histoire naturelle and chair of comparative anatomy, by means of endocranial casts, or access to the endocranial face of fossils. These prints are left by the venous vascularization of the meninges. In his book, *Anatomie et évolution des veines méningées chez les hommes fossiles* (Anatomy and Evolution of Meningeal Veins in Fossil Humans), Saban (1984) explores vascular prints, openings and diploic canals, and their relationships with the vessels linked with the meningeal envelopes (arteries, veins, venous sinuses and blood pools) (see Figure 2.1). This book, presented as a reference for human paleontology (Heim 1986), served as a guiding thread for us, qualifying the phylogenetic interpretations modified and enriched by data accumulated in paleoanthropology (Broadfield et al. 2010; Grimaud Hervé and Lordkipanidze 2010; Bruner 2017).

His observations positioned Neanderthals relative to the different African, European and Asian lineages. His conclusion surprised Jean-Louis Heim (1986), a specialist in Neanderthal bone growth: their encephalization would have been followed by a depletion of venous drainage:

> This vascular system regulating cerebral blood drainage demonstrates the progressive complication of the network, the only one that can be known in fossil forms, thus reflecting the degree of vascularization of the encephalon. The various morphological stages are inscribed by specific models, generally in relation to the increase in encephalic volume. During the course of this evolution we witness the establishment of this vascular system in *Australopithecus*, then its complication from *Pithecanthropus*[1] onwards, through the multiplication of the ramifications and their anastomoses, to achieve in modern humans a veritable vascular network covering the entire parietal surface. Only Neanderthals consistently show the presence of the anterior great vein (sinus of Breschet), which, in our current forms, appears to be an occasional resurgence of an archaic characteristic (Saban 2007).

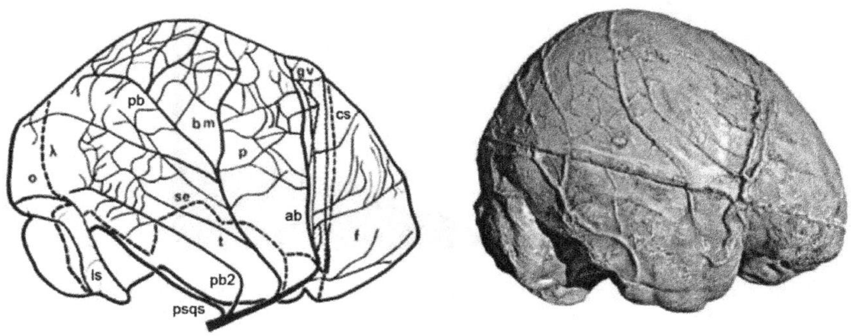

a) b)

Figure 2.1. *Network of middle meningeal veins in modern humans. For a color version of this figure see www.iste.co.uk/dambricourtmalasse/futureofsapiens.zip*

COMMENT ON FIGURE 2.1.– a) Drawing in the right lateral view, ab: anterior branch, mb: middle branch, pb: posterior branch, pb2: secondary posterior branch, f: frontal region, gv: anterior great vein, λ: lambdoid suture, o: occipital region, p: parietal

---

1 Pithecanthropes are the *Homo erectus* of Asia considered for a time as the ancestors of all other Eurasian and African lineages.

region, cs: coronal suture, ss: scaly suture, ls: lateral sinus, psqs: petrosquamous sinus, t: temporal region. b) Endocranial casting of a present-day man (according to Saban 1987) ((a) Collections of the Institut de paléontologie humaine, (b) photo A. Dambricourt Malassé. Figure in Dambricourt Malassé 2015.)

This book was supplemented by a synthesis entitled "*Aux sources du language articulé*" (Saban 1993). We are therefore interested in these observations in order to verify whether recent works have invalidated them or whether it is relevant to establish physiological cardiovascular correlations with the particular postural development of Neanderthals. Indeed, it is distinguished from older hominins by a collapsing of its structures from the skull base, with lordosis disappearing, suggesting a decrease in the tensions of the connective tissue lining the neural tube (the primitive meninx), especially during the migration of cells from the neural crest and the closure of the fourth ventricle at the apex of the dorsal chord. The hypothesis put forward is a decrease in embryonic metabolism (see Chapter 1).

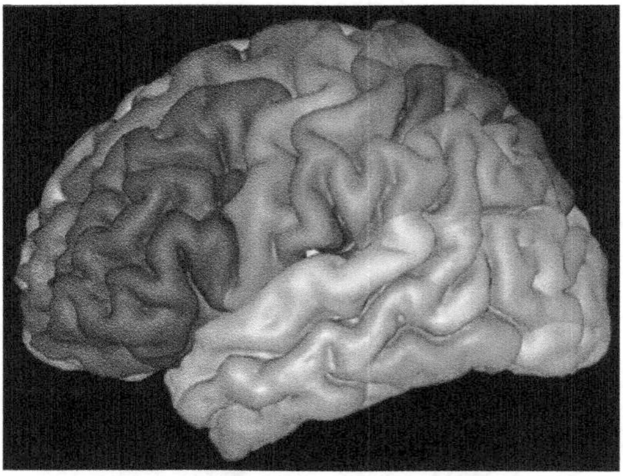

Figure 2.2. *Cerebral gyri in the left lateral view (Cantin et al. 2011). For a color version of this figure see www.iste.co.uk/dambricourtmalasse/futureofsapiens.zip*

COMMENT ON FIGURE 2.2.– The frontal lobe with the precentral gyrus (pink) and the prefrontal cortex composed of the gyri: superior frontal (red), middle frontal (burgundy) and inferior frontal (brown); the parietal lobe with gyri; postcentral (blue), superior parietal (anthracite gray) and inferior parietal (purple); the temporal lobe with gyri: superior temporal (yellow), middle temporal (yellowish gray) and inferior temporal (greenish yellow); the occipital lobe with gyri: superior occipital (spring green), middle occipital (light green) and inferior occipital (pale green)[2].

2 See: www.radioanatomie.com.

Figure 2.2 shows the different convolutions, or gyri, of the neocortex of modern humans. For a review of publications on endocranial prints of fossil hominins, we refer to the study of Guercy 1 (Ardèche, France) attributed to a primitive Neanderthal (Richards et al. 2022). The authors of this latest study point out that the oldest Neanderthals are distinguished from pre-Neanderthals by a pronounced development of the prefrontal cortex, the precentral and postcentral gyri, the inferior parietal lobule and the frontoparietal operculum (a lobule located on the inner face of the hemispheres), while Neanderthals continued these developments, even more significant for the parietal and temporal gyri. Roger Saban therefore compared the middle meningeal vascularizations of extinct hominins, and their growth, with those of modern humans. Neanderthals differ from older species, with less cranial capacity and less-complex cerebral lobes, by a reduction in the ramescence of the three branches and their anastomoses, as well as by the conservation of the petrosquamous sinus and the anterior great vein of Galen. This vein, called "sinus of Breschet", borders the small wing of the sphenoid, yet there exists a parietal print of the great vein (gv), which is rarely mentioned (Saban 2007) (see Figures 2.3 and 2.4).

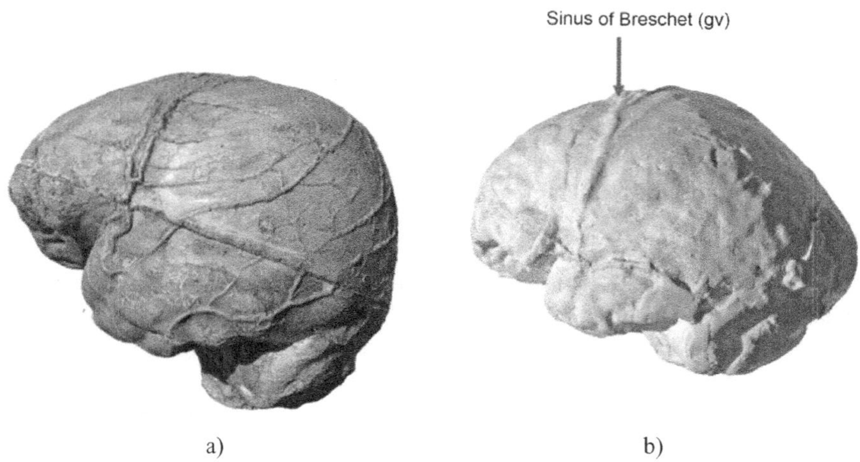

Figure 2.3. *Endocranial casts showing the prints of the middle meningeal veins in a modern human (a), and an adolescent Neanderthal (Teshik Tash, Uzbekistan) (b) with the sinus of Breschet (gv: great vein) (collections of the Institut de paléontologie humaine, photos and copyright: A. Dambricourt Malassé). For a color version of this figure see www.iste.co.uk/dambricourtmalasse/futureofsapiens.zip*

The vein of Galen is formed in the fetus and replaces the median vein of Markowski, an embryonic vessel of the primitive meninx that forms with the prosencephalon (the future cerebral hemisphere). The parietal print of a great vein is the persistence of the embryonic vein of Markowski. In modern humans, its persistence is an anomaly linked to an aneurysmal malformation (a dilation) of the vein of Galen. We will see how this dilation of an embryonic vessel forms part of the convergence of observations signaling a change in embryonic metabolism.

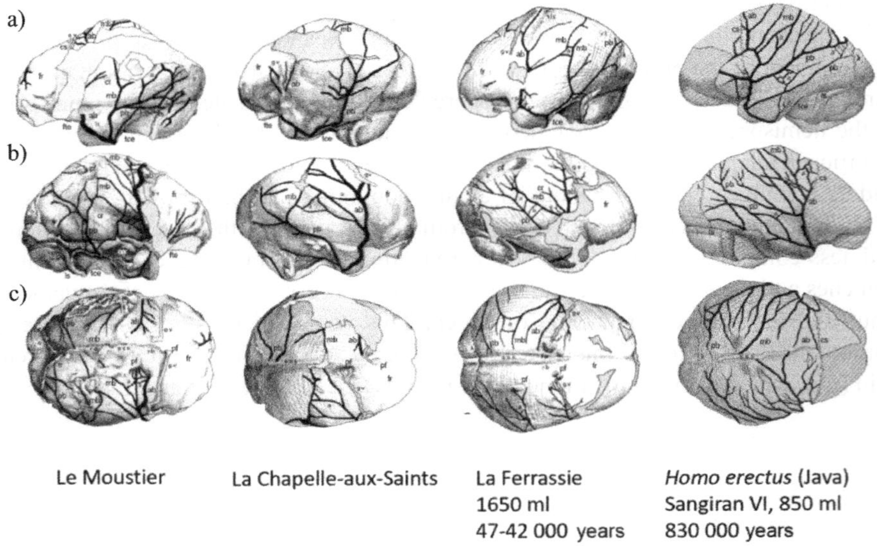

Figure 2.4. *Network of the middle meningeal veins in three adult Neanderthals and one adult* Homo erectus

COMMENT ON FIGURE 2.4.– a) Left lateral view, b) right lateral view, c) top view: a: anastomosis, ab: anterior branch, mb: middle branch, pb: posterior branch, fte: fronto-temporal excavation, tce: temporo-cerebellar excavation, pf: Pacchioni's fossa, gv: anterior great vein, λ: lambdoid suture, cr: central parietal region, ss: scaly suture, fr: frontal region, lr: lambdatic region, cs: coronal suture, ls: lateral sinus, psqs: petrosquamous sinus, sss: superior sagittal sinus (different scales) (modified from Saban (1987); figure in Dambricourt Malassé (2015)).

The identification of vascular prints in a sample of modern humans can be subject to considerable variation between observers (Labra et al. 2023). We take these remarks into account while acknowledging Roger Saban's rare experience in comparative anatomy, in that it covers all primates (Saban 1986). Furthermore, one of us was confronted with this reconstruction of the middle meningeal vascularization for the skull of Skhul I (140 ka, Israel) (Bouvier et al. 2025), the oldest child endowed with *sapiens* characteristics, or anatomically modern sensu stricto, i.e. from the earliest stages of cephalic embryogenesis. Its meningeal vascularization was poorly developed for its age. The study of ultra-high-resolution computed tomography conducted by the Shmunis Family Anthropological Institute of Tel Aviv University shows no differences with the interpretations of McCown and Keith, published in 1937–1939. With a few rare exceptions, Saban's conclusions are correct:

> This four- to five-year-old child shows a great delay in the development of this vascular network compared to a present-day child. It seems barely comparable to what we know of a one-year-old child, who already presents more anastomosis. This delay seems to be in line with that of the Neanderthal child of La Ferrassie and the Gibraltar child of comparable age (Saban 1984).

The skull, including the mandible, presents a mixture of *Homo sapiens* and Neanderthal morphogenetic characteristics, the postural and physiological interpretation requiring further investigation.

2.2.2. *The diploic and emissary venous system*

The human brain uses 20–25% of the body's energy resources, while other current primates consume only 8–10%, and other mammals 3–5% (Leonard et al. 2007; Squire 2012; Wang et al. 2016). The vascularization of the spongy tissue, which separates the inner and outer cortical layers of the vault bones, or diploë (see Figure 2.5(a)), is thought to play a role in the thermoregulation of the voluminous brains of the genus *Homo*. This network is connected to the sinuses of the dura mater at four levels of the neurocranium: frontal, anterior temporal, posterior temporal and occipital. The development of the diploic veins would therefore be a neoformation of the vascular system necessary for cooling large brains such as that of the Neanderthal (Bruner et al. 2011).

Rangel-de Lazaro et al. (2016) gave a quantitative description of the frontal, parietal and occipital diploic networks of three Neanderthal skulls: Saccopastore 1 (Italy), Spy 1 and Spy 2 (Belgium), compared to 20 modern human skulls. The vascular network of the three Neanderthals is situated within the lower range of modern variation. A similar study was undertaken on Spy 1 and on four other Neanderthals: Spy 10 (Belgium), La Chapelle-aux-Saints, La Ferrassie 1 and La Quina H5 (France) in order to test the thermoregulation hypothesis (Hui and Balzeau 2023a, 2023b). The results show an amplitude of variations that is too great in modern humans to establish taxonomic distinctions. In contrast, the way in which the diploë veins communicate with the meningeal and pericranial vessels and the dural sinuses presents clear differences in the connections surrounding the frontal sinuses, parietal foramina and the asteric region.

Emissary drainage, meanwhile, is an auxiliary substitution pathway composed of low-flow anastomotic veins connecting the venous sinuses of the skull base to the extracranial venous system. It is studied thanks to the emissary foramina of the skull (see Figure 2.5(b)). The orifices of the emissary veins are visible on the fossil skulls: the condylar (or hypoglossal) canal, which allows the passage of the occipital emissary vein in order for it to connect to the suboccipital venous plexus, the occipital sinus and the sigmoid sinus; the parietal foramen, which corresponds to the passage of the parietal emissary vein in order to reach the superior sagittal sinus; and the mastoid foramen, which makes the passage between an emissary vein behind the mastoid process and the sigmoid sinus, as well as the suboccipital venous plexus.

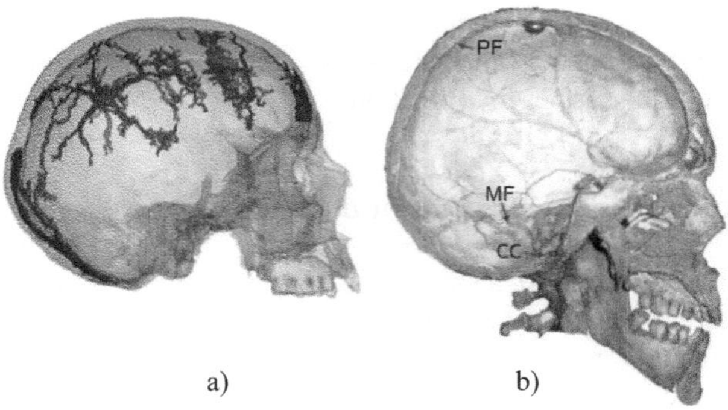

Figure 2.5. *Vascularization of the endocranium of a modern* Homo sapiens*: a) three-dimensional reconstruction of diploic venous canals; b) vascular prints and orifices of emissaries (according to Píšová et al. (2017). (a) from the collection of the Institut de paléontologie humaine; (b) photo and copyright A. Dambricourt Malassé. Figure in Dambricourt Malassé 2023). CC = condylar canal, MF = mastoid foramen, PF = parietal foramen. For a color version of this figure see www.iste.co.uk/dambricourtmalasse/futureofsapiens.zip*

Their number increased with the complexification of the encephalon, and the hypothesis arose of a link with temperature. Indeed, as blood flows from the outside to the inside of the skull to cool the brain area, the venous drainage flow should follow. Pélissier-Hermittea (2006) tested this hypothesis in the lineage of Neanderthals (2 pre-Neanderthals and 20 Neanderthals). The results show that the occurrence frequencies of emissary foramina are always higher in modern humans relative to fossil species, with the exception of Neanderthals who lived in glacial climatic conditions. According to the author, the increase in emissary drainage, both quantitative and qualitative, follows the phylogenetic time curve (evolution) more than the thermal curves of glacial and interglacial climates.

2.3. The state of current knowledge versus archaic, and vice versa

It seems very interesting to us here to question the comparison of the vascularization of the central nervous system between Neanderthals and modern humans because while their brains have practically the same volume, they have a totally different shape (see Figures 2.6 and 2.7).

Figure 2.6. *Endocranial castings (same scale) in the left lateral view showing differences in the shape of adult cerebral hemispheres in a fossil* Homo sapiens *(Předmostí 3, 25–27 ka, Czech Republic) (a), and a Neanderthal (La Chapelle-aux-Saints, 45 ka, France) (b) (collections of the* Institut de paléontologie humaine, *photo and copyright: A. Dambricourt Malassé). For a color version of this figure see www.iste.co.uk/dambricourtmalasse/futureofsapiens.zip*

The cortical distribution specific to each and their large morphological differences are associated with the upper vault, the upper parietal part constituting a "hot zone" (see Figures 2.4 and 2.6). According to the study by Bruner et al. (2014), and in line with a temperature distribution model based on endocranial geometry, Neanderthals and modern humans present different patterns of heat dissipation. In particular, the upper parietal area of Neanderthals is very flat and closer to the "geometric heart of the brain", according to the expression of Píšová et al. (2017), which represents a large surface area for dissipation. In contrast, in modern humans, this area is included in the deeper cerebral volume as a result of the volumetric development of the parietal lobes (see Figure 2.7).

According to Bertolizio et al. (2011) and Wang et al. (2016), heat production and thermal regulation are dependent on several factors, the crucial role of which is attributed specifically to the vascular system. In modern humans, anastomoses between arterial and venous vessels serve to regulate temperature. They are found in the most peripheral regions of the human body, such as the hands. Thus, when it is cold, these anastomoses will open and the capillary network will be less irrigated.

This vasoconstriction process thus makes it possible to regulate the temperature of the vital organs to the detriment of the periphery. It is the volume of the encephalon that is responsible for this heat production, and the cortical surface is responsible for this heat dissipation in the brain. Minor changes in shape are also noted that may influence heat distribution and the specific cortical areas (Rival et al. 2009).

Figure 2.7. *Endocranial castings (same scale) in top view showing volumetric differences in adult cerebral lobes in a fossil Homo sapiens (Předmostí 3, 25–27 ka, Czech Republic) (a), and a Neanderthal (La Chapelle-aux-Saints, 45 ka, France) (b) (collections of the Institut de paléontologie humaine, photo and copyright: A. Dambricourt Malassé). For a color version of this figure see www.iste.co.uk/dambricourtmalasse/futureofsapiens.zip*

Saban has shown that we are witnessing, over the course of the biological evolution of hominins, the establishment of a more rudimentary vascular system, first of all in Australopithecus, then its increasing complexification through the multiplication of the ramifications and their anastomoses, to reach in anatomically modern humans a vascular network that covers the entire parietal surface. Only Neanderthals consistently show the presence of the anterior great vein (sinus of Breschet or sphenoparietal sinus), which, in our current forms, appears to be "an occasional resurgence of an archaic characteristic".

Even if the prints of the external surface of the Neanderthal encephalon present similarities with those of modern humans, for example, with the presence of Broca and Wernicke areas, in contrast, the meningeal vascular topography thus indicates differences, some even bringing them closer to certain *Homo erectus* with, in particular, a particularly developed sinus of Breschet, as on Sangiran II of Java (Saban 1984). Still according to the studies of Saban (1982, 1984) (see Figure 2.8), Neanderthals thus possessed a parietal vascular network much less developed than modern humans, with few anastomoses present, which strongly suggests that blood vessels can adapt to physiological or functional changes over the course of evolution (Píšová et al. 2017).

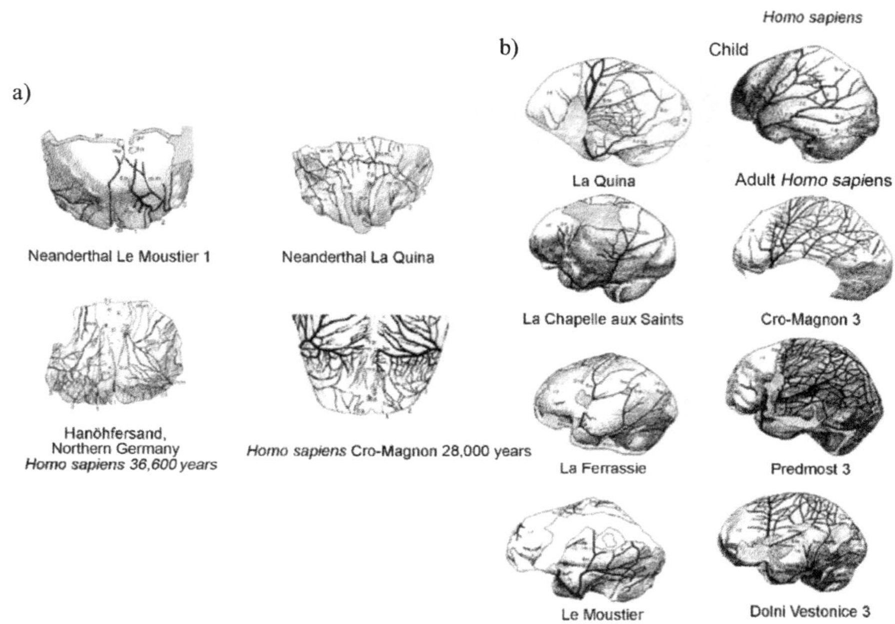

Figure 2.8. *Middle meningeal vascularization of Neanderthals and* Sapiens*: a) the Neanderthal anterior network is poorly developed compared to Sapiens; b) comparison between adult Neanderthals and fossil* Homo sapiens, *and with a present-day 1-year-old child (modified from Saban (1982, 1984), figure in Dambricourt Malassé 2015)*

The volume increase of the parietal lobes of Neanderthals was therefore not accompanied by a complexification of the anastomoses. On the contrary, it has lessened. Does this mean that the metabolism of the parietal lobes did not require cooling? Or was the drainage of the diploic veins, with the sinuses of the dura mater not ensured at the parietal level by the sinus of Breschet from the fetal period?

2.4. Current data versus postnatal development and aging

It is well known nowadays that normal brain development in children is essential to the acquisition of optimal cerebral functions in adults. Similarly, the possible constraints to which the development of the human brain is subjected have implicit importance for understanding the constraints applicable to the higher faculties of the brain. Thus, for modern humans, the process of complexification of the brain increases with the development of the encephalon via evolutionary and sensory awakening states that stimulate the senses and motor skills (see Figure 2.9).

Figure 2.9. *Learning how to walk and locomotor autonomy. For a color version of this figure see www.iste.co.uk/dambricourtmalasse/futureofsapiens.zip*

The brain of modern humans accounts for an average of 2% of body weight and consumes, for a resting individual, nearly 20% of oxygen breathed and uses 20–25% of total body energy, compared to a maximum of 5% in other mammals (Leonard and Robertson 1997). Moreover, our brain is highly sensitive to the variations of certain metabolites and especially to the blood sugar level, and thus requires energy permanently and continuously: fuel (about 100 mg/min of glucose) and combustive: oxygen (Bourre 2004).

The specific morphology of meningeal vessels within our species and the persistence of these branches in modern adults suggest active roles for these vascular systems, which are still poorly understood today. Blood flow is low in a resting adult, particularly in the main branches of the meningeal canals (Bruner et al. 2011). This suggests that their role is more fundamental during the early stages of development, or that these vessels function with an increase in heat, particularly during a state of physiological stress (physical and/or sporting activity, pathology, illness or mood states).

Píšová et al. (2017) hypothesized that the particular development of these vessels in our species is the consequence of physiological and evolutionary processes that have targeted cerebral vessels and had a secondary effect on the meningeal and diploic system, through common growth factors.

We can draw parallels between this complexification mechanism and the period of aging and its pathologies; with the case of Binswanger encephalopathy, on the one hand, or Alzheimer's, on the other.

Binswanger's encephalopathy is a neurological disease characterized by arteriopathic or vascular dementia, generally occurring in individuals with high blood pressure and corresponding to a variety of dementia. This neurological condition appears in individuals approximately 60 years of age and older, who also show signs of arteriosclerosis leading to high blood pressure with damage to the large vessels due to atheromatous deposits. It is a neurological condition resulting from vascular lesions due to infarcts affecting the white matter of the cerebral hemispheres. The patient has difficulty walking before the actual onset of dementia.

For Alzheimer's disease, metabolic decreases in the posterior parietal cortex have been described as early periods of the disease (Jacobs 2012), but other pathologies classified as degenerative are the expression of a consequence of microstroke (Binswanger, subcortical dementias: signs of leukoaraiosis, leukopathy, Parkinson's, Lewy body disease, etc.).

In the history of life, according to Crawford (2002), the dietary acquisition of one molecule, DHA (Docosahexaenoic acid, also called cervonic acid, which is an omega-3 polyunsaturated fatty acid), plays the role of pacemaker for metabolic activities and has largely contributed to the evolution of the human brain.

In conclusion, it can thus be assumed that the diet of Neanderthals may have contributed, in addition to its genetic and morphogenetic characteristics, to the vascular development of the brain, whose morphology is very much specific compared to modern humans.

2.4.1. *The cardiovascular system and postural development of the Neanderthal*

Neanderthal morphogenesis was faster than that of modern humans and different before birth; the morphogenetic growth patterns are not similar. These observations have been known for the mandible and the internal base of the skull for nearly 40 years, regularly confirmed by advances in medical imaging, with tomography applied to fossils, consolidating the first biological level concerned by the processes of hominization and their diversities, namely, the embryonic body plan. The characteristics shared by the different lineages, each having its own derived characteristics, are described as plesiomorphic. The body plan of Neanderthals remains within the limits of the oldest known species of the genus *Homo*, whereas the modern human is its prolonged and complex embryonic reorganization, with the development of the nervous system (see the Introduction and Chapter 1 of the current book).

2.4.1.1. *The rib cage*

The rib cage of Neanderthal children is different from that of modern humans before birth; it would represent the plesiomorphic condition shared with *Homo erectus* (Garcia-Martínez et al. 2020). Bell-shaped, not barrel-shaped, its great lower width matches the space between the two iliac wings of the pelvis. This enlargement of an ancestral organization is interpreted as a likely adaptation of the kidneys and liver to glacial winter periods, following a depletion of the fat- and carbohydrate-rich resources provided until then by big game. These deficiencies would have been compensated for by a greater consumption of protein as an energy source, leading to an increase in the activity, weight and probably volume of the liver and the urinary system (Ben-Dor et al. 2016). The large width would also correspond to a developed heart and lungs. The shape of the rib cage suggests a larger diaphragm than that of *Homo sapiens*, allowing for maximum air storage; these volumes were about 20% greater than the corresponding volumes of modern humans of the same size and sex. The respiratory function of Neanderthals was therefore different from modern humans and their capacity was greater (Garcia-Martínez et al. 2018). Lastly, maternal diet may have had an impact on the development of the fetus (Weaver et al. 2016) and been detrimental to them. Paleontologist Jean Chaline (2003) proposed the hypothesis of eclampsia to understand their extinction by referring to Camilleri:

> It is a dysfunction of the placenta, which releases into the maternal circulation high concentrations of molecules that become toxic, particularly to the kidneys and the blood-vessel walls. Associated with high blood pressure and the appearance of protein in the urine, this placental dysfunction also affects the baby, who will often experience growth retardation in utero (Camilleri 2023).

2.4.1.2. *The cardiovascular system of great apes and of modern and fossil humans*

The research by Drane and his collaborators on great apes living in reserves suggests that the human heart has adapted to verticality, consecutive body movements and the volumetric development of the brain (Curry et al. 2024). These studies of the cardiovascular system compared the structure of the heart, its function, its size and how the muscle contracts, twists and turns. The high cardiac volume values appeared to be "outliers" and are interpreted as a metabolic demand for blood supply, much greater for muscles and the brain in humans than in great apes, with a higher number of heartbeats per minute for the whole body. Aortic flow during systole is the product of the amount of blood pumped by the left ventricle in a single heartbeat (systolic volume) and the number of heartbeats per minute (heart rate). Drane discovered that the left ventricle, which is the heart's largest pumping chamber, contains muscle bundles arranged in a network in all great apes, called "trabeculations", whereas the human ventricle has a smooth and particularly soft wall at its apex (the tip of the heart). The researcher discovered a second striking difference in human function when muscle thickens, twists, turns and lengthens. Twisting and rotation are much greater at the apex during contraction of the ventricles, which have the fewest trabeculae. The trabeculated myocardium common to modern great apes is inherited from common ancestors, and was therefore also that of the semi-upright great apes, ancestors of hominins with a verticalized axial skeleton. When did trabecularization begin to diminish during the course of hominin evolution?

Ríos et al. (2023) sought to evaluate the relationship between cardiac output and energy expenditure over the course of hominin evolution, through an initial study of the descending aortic print visible on the vertebral bodies from the T5 thoracic to the L2 lumbar vertebrae. This print can be observed by an asymmetry of the vertebral body, with the left face being flatter. This deformation is non-existent in great apes. The authors also considered aortic root diameter as an indirect measure of cardiac output. Humans have a larger body mass-adjusted diameter than gorillas and chimpanzees, with a pronounced increase during the period of brain growth. The shape of the rib cage is also taken into consideration. It is smaller and flatter in humans than in gorillas, where it is wider and deeper; these morphological differences could imply a tighter enclosure of the rib cage contents in humans, thus contributing to the development of the aortic print on the vertebrae. The authors conclude that the aortic print is a skeletal signal of a large aorta, in a reduced rib cage, with a high metabolism requiring a high cardiac output per unit of body mass.

Neanderthals present a slightly deeper thorax than modern humans, but the vertebra of El Sidrón (Spain) presents a clear aortic print. In contrast, the teenage *Homo ergaster* of the Kenyan River Nariokotome (KNM-WT 15,000), dated 1.5–1.6 Ma, already had a thorax that was considerably deeper anteroposteriorly than modern humans, but its vertebrae do not show a print of the aorta. Its brain was small (880 cm^3) relative to its height (1.65 m). The endocranial casting shows prints of the well-preserved meningeal vascular network, with both anterior and posterior branches (Begun and Walker 1993); however, their ramifications are less numerous than those of *Homo habilis* of 1.9 Ma, KNM ER 1813, and even less in comparison with KNM ER 1470, with its communications between the two branches (Saban 1984). The shape of the thorax and the absence of an aortic print, despite its size and long legs, were met with surprise, because the anatomy of the ancient African hominins is considered as an adaptive response to endurance racing, which requires an increase in metabolism and therefore cardiac output compared to the great ape.

2.4.2. *Embryogenesis of the cardiovascular system and acquisition of axial verticality of the embryo*

Jensen et al. (2016) called into question the hypothesis that abnormal hypertrabeculation of the left ventricle in humans, still poorly understood, is the preservation of the embryonic state of the ventricle of ectothermic (cold-blooded) vertebrates.

2.4.2.1. *Embryogenesis*

The authors studied a series of embryos, both human and from nine other species, as well as adult hearts from 15 species. Early in development, vertebrate embryos have a ventricle formed of 80% trabeculations, but only the ectothermic species have retained them throughout development. In endothermic (warm-blooded) species, including humans, the heart changes with fetal and postnatal development, with a significant reduction in small trabeculations, generally under 30%. In humans, hypertrabeculated left ventricles are, nevertheless, much less so than at the embryonic stage (15–40% volume). The authors conclude that hypertrabecularization is not the preservation of the embryonic state, nor the homologue of adult ectothermic species. Their conclusions relating to the common similarity with the heart of great apes and its function, however, have been reviewed since the works of Drane et al. (2019) and Curry et al. (2024).

We can conclude that the reduction of embryonic hypertrabecularization in modern humans is considerably greater during fetal development than in great apes. Thus, if the osteological signals of the development of the aorta are absent on the skeleton of Nariokotome, the hypothesis of a trabecularization still pronounced in these hominins with a small brain, with a relatively less-developed middle meningeal vascularization than *Homo habilis,* is conceivable.

NOTE.– Fetal reduction in trabecularization most likely followed encephalization, as evidenced by the aortic print of the Neanderthal El Sidrón.

The embryonic and dynamic origin of the axial straightening, which corresponds to the path of the dorsal chord, has been demonstrated with the dorsoventral rotation of the post-sphenoid and basi-occipital synchondroses. These movements are most likely generated by the tensions of the neural crest cells during their migration toward the ventral pole of rhombomeres 4, 5 and 6, under the control of Hox genes

(Dambricourt Malassé et al. 2025b). The sphenoidal amplitude of rotation in modern humans is greater than that of any fossil species of the genus *Homo*. All have the same amplitude in common, however, and the Neanderthals did not exceed it. The regulation of Hox genes along the rostro-caudal axis of the embryo was therefore modified with *Sapiens*. The amplitude of rotation of the modern chondrocranium shows that this modification occurred, at the cephalic pole, in the direction of a prolongation of migration. In other words, the reception of signals indicating that migration must stop occurs later in modern humans. The regulation of these signals involved in the morphogenesis of axial support tissues, with the skull, vertebrae and rib cage, and appendicular support tissues with the girdles, limbs and extremities, was therefore different in the Neanderthal and probably ceased earlier.

2.4.2.2. *The postnatal posture and cardiovascular system of the Neanderthal*

Neanderthal skeletal anatomy is in no way comparable to modern anatomy, from the labyrinth to its thorax, from the pectoral girdle to the pelvic girdle, to the extremities of the limbs. A study of the bones of the foot focused on its morphology, the talus in particular (Sorrentino et al. 2021). The authors conclude, in turn, that "ancestral retentions" were associated with a trabecular bone volume fraction that suggests "an intense and consistently-oriented locomotor load". Neanderthals also present differences in the ankle joint (talocrural or tibiotarsal) compared to *Homo sapiens*, attributable to a locomotor "dissimilarity". The talonavicular joint, located on the dorsal and internal edge of the foot, contributes to the adaptation and stability of the foot on unstable terrain. In the Neanderthal, it presents a mixture of ancestral and functional traits that are added to a derived subtalar joint (between the talus and the calcaneus), all suggesting a predisposition for a foot in pronation during the stance phase. The authors suggest that a greater body mass and/or higher mechanical stress led to their usual pronated foot posture.

The embryonic and dynamic postural approach provides an understanding of these differences, which thus appear as the physiological and functional postnatal consequences. It is supported by the conclusions of Weaver et al. (2016):

> Neonatal features could not have developed in response to a mechanical loading associated with adult/older-subadult behaviors (e.g. hunting), so they must either have an evolutionary explanation or result from environmental influences on the fetus (e.g. maternal diet) (Weaver et al. 2016).

At birth, the Neanderthal was already in an extended basi-craniosacral dynamic, with a skeletal morphology characterized by a wide thoracic cavity. Its psychomotor balance will be acquired according to the body's axis of gravity and a tensioning of the musculoskeletal framework from the head to the soles of the feet, with a gravitational load line that is different from the modern human. In our species, this load line passes in front of the C7/T1 vertebrae, behind T12/L1 and the tangent of the femoral axis, with the sacrum positioned in anteversion. In Neanderthals, the dynamic is reversed. The "collapsed" posture and the shifted axis of gravity impose a shift in the weight load that affects the ankles and soles of the feet. The gravity load line is located well in front of the thoracolumbar vertebral column (see details in Dambricourt Malassé et al. 2025b), modifying the postural constraints with a retroversion of the sacrum and knees in semi-flexion. This is followed by an internal rotation of the tibial compartments that drives the ankles, tarsals and metatarsals in a ground-gripping movement in pronation. This is also demonstrated in modern humans with hip dysplasia and knee joint malformations (Hadjouis 2006).

An evolution of the embryonic metabolism, in terms of depletion of the energy intake of the maternal body, seems very likely with compensations of the regulation of the embryonic and fetal cardiovascular system. The middle meningeal vascularization is poorly developed, but it has retained the sinus of Markowski, whereas the latter disappears during fetal development in modern humans and therefore very likely disappeared in the oldest known species of the genus *Homo*. This raises the question of whether the formation of the heart and therefore detrabecularization has also been slowed down, reducing the number of heart pulsations, while the brain and cerebellum needed a significant blood supply for their development and activities. The volume of blood contained in the ventricles and passing through the aorta would have been greater per heartbeat; the size of the ventricles, the diameter of the aorta and the blood vessels of the meninges would have increased; the sinus of Markowski would have been preserved; and the anterior great vein of Galen dilated to ensure the drainage of a blood supply necessary for the metabolism of the encephalon in formation.

2.5. Conclusion

Even if, from the 22nd day for the modern human embryo, the first heartbeats are already taking place, like all the other equally crucial functions that will follow from birth for the functional autonomy of this future individual, nevertheless from this very early period of development, there follows a whole multitude of exponential growth actions that will stop only after puberty. The role of the heart from its very activity as a cardiac tube already possesses that of the pacemaker of the human body, and all the major functions depend on its network, which, connected by the vascular system, allows its constant nourishing supply. Can we therefore conceive in the Neanderthal, throughout its life history, of functional physiological correlations for the heart, and lungs developed to compensate for a slowed metabolism from embryogenesis, requiring a higher oxygen level in the body than in its contemporary *Homo sapiens* for the same unit of time?

2.6. References

Begun, D. and Walker, A. (1993). The endocast. In *The Nariokotome Homo Erectus Skeleton*, Walker, A. and Leakey, R. (eds). Harvard University Press, Cambridge, 326–358.

Ben-Dor, M., Gopher, A., Barkai, R. (2016). Neandertals' large lower thorax may represent adaptation to high protein diet. *Am. J. Phys. Anthropol.*, 160(3), 367–378. doi: 10.1002/ajpa.22981.

Bertolizio, G., Mason, L., Bissonnette, B. (2011). Brain temperature: Heat production, elimination and clinical relevance. *Pediatr. Anesthesia*, 21, 347–358. doi: 10.1111/j.1460-9592.2011.03542.x.

Bourre, J.M. (2004). The role of nutritional factors on the structure and function of the brain: An update on dietary requirements. *Rev. Neurol.*, 160(8/9), 767–792.

Bouvier, B., Dambricourt Malassé, A., Otte, M., Hershkovitz, I. (2025). A new analysis of the neurocranium and mandible of the Skhūl I child: Taxonomic and cultural implications. *L'Anthropologie*, 4.

Broadfield, D., Yuan, M., Schick, K.D., Toth, N. (eds) (2010). *The Human Rain Evolving: Paleoneurological Studies in Honor of Ralph L. Holloway*. Stone Age Institute Press, Gosport.

Bruner, E. (2017). The fossil evidence of human brain evolution. In *Evolution of Nervous Systems*, 2nd edition, Kaas, J. (ed.). Elsevier, Oxford, 63–92.

Bruner, E., Mantini, S., Musso, F., De La Cuétara, J.M., Ripani, M., Sherkat, S. (2011). The evolution of the meningeal vascular system in the human genus: From brain shape to thermoregulation. *Am. J. Hum. Biol.*, 23(11), 35–43. doi: 10.1002/ajhb.21123.

Bruner, E., de la Cuétara, J.M., Masters, M., Amano, H., Ogihara, N. (2014). Functional craniology and brain evolution: From paleontology to biomedicine. *Front. Neuroanat.*, 8, 1–15.

Camilleri, C. (2023). Grossesse santé contre la pré-éclampsie. *Association Grossesse Santé* [Online]. Available at: https://grossesse-sante.org/.

Cantin, S., Mendoza, C., Grand, S., Bessou, P., Lefournier, V., Eker, O., Tahon, F., Bing, F., Cuvinciuc, V., Jiang, Z. et al. (2011). Imagerie par résonance magnétique cranioencéphalique normale : étude en coupes frontales, sagittales et axiales. *EMC consulte, Radiologie et imagerie médicale (musculosquelettique - neurologique - maxillofaciale)*, 30-806-A-10.

Chaline, J. (2003). Increased cranial capacity in hominid evolution and preeclampsia. *Journal of Reproductive Immunology*, 59(2), 137-152.

Chapon Sao, C., Tudryn, A., Dambricourt Malassé, A., Moigne, A.M., Gargani, J., Singh, M., Abdessadok, S., Voinchet, P., Cauche, D., Karir, B. et al. (2024). Magnetostratigraphy of the Pliocene Masol Formation, Siwalik Frontal Range, India: Implications for the age of intentional cut-marked fossil bones. *J. Asian Earth Sci.*, 259. doi: 10.1016/j.jseaes.2023.105884.

Crawford, M. (2002). Cerebral evolution. *Nutrition and Health*, 16, 29–34. doi: 10.1177/026010600201600108.

Curry, B.A., Drane, A.L., Atencia, R., Feltrer, Y., Calvi, T., Milnes, E.L., Moittié, S., Weigold, A., Knauf-Witzens, T., Sawung Kusuma, A. et al. (2024). Left ventricular trabeculation in Hominidae: Divergence of the human cardiac phenotype. *Communications Biology*, 7(1), 682. doi: 10.1038/s42003-024-06280-9.

Dambricourt Malassé, A. (2015). La reconstitution du développement psychomoteur *d'Homo neanderthalensis* : une nécessité pluridisciplinaire. Courses of the Doctoral School ED227 Diversities, Origins, Natures, Muséum national d'Histoire naturelle and Sorbonne Université, Institut de Paléontologie Humaine, Paris.

Dambricourt Malassé, A. (2023). Origines et avenir de notre verticalité. Que nous révèlent les collections de fossiles ? In *Journées européennes du patrimoine*. Institut de Paléontologie Humaine, Paris.

Dambricourt Malassé, A. and Cauche, D. (2022). Des traces de boucherie aux outils lithiques de Masol (Pliocène supérieur, Inde du Nord). Une approche psychomotrice de l'origine et de l'identité des hominiens. *L'Anthropologie*, 4. doi: 10.1016/j.anthro.2022.103070.

Dambricourt Malassé, A., Tudryn, A., Gargani, J., Moigne, A.-M., Chapon-Sao, C., Abdessadok, S., Singh, M., Cauche, D., Voinchet, P., Karir, B. et al. (2025c). Masol (India, Punjab): Horses (*Equus*) and hominins in the sub-Himalayan floodplain as early as 2.95 Ma. *L'Anthropologie*, 129(4), 103392. doi: 10.1016/j.anthro.2025.103392..

Drane, A.L., Atencia, R., Cooper, S.-M., Rodriguez, P., Sanchez, C., Simcox, S., Feltrer, Y., Peck, B., Eng, J., Moittie, S. (2019). Cardiac structure and function characterized across age groups and between sexes in healthy wild-born captive chimpanzees (*Pan troglodytes*) living in sanctuaries. *Am. J. Veterinary Res.*, 80, 547–557. doi: 10.2460/ajvr.80.6.547.

García-Martínez, D., Torres-Tamayo, N., Torres-Sánchez, I., García-Río, F., Rosas, A., Bastir, M. (2018). Ribcage measurements indicate greater lung capacity in Neanderthals and Lower Pleistocene hominins compared to modern humans. *Commun. Biol.*, 16(1), 117. doi: 10.1038/s42003-018-0125-4.

García-Martínez, D., Bastir, M., Gómez-Olivencia, A., Maureille, B., Golovanova, L., Doronichev, V., Akazawa, T., Kondo, O., Ishida, H., Gascho, D. et al. (2020). Early development of the Neanderthal ribcage reveals a different body shape at birth compared to modern humans. *Sci. Adv.*, 6, eabb4377. doi: 10.1126/sciadv.abb4377.

Grimaud Hervé, D. and Lordkipanidze, D. (2010). The fossil hominid brains of Dmanisi: D 2280 and D 2282. In *The Human Brain Evolving: Paleoneurological Studies in Honor of Ralph L. Holloway*, Broadfield, D., Yuan, M., Schick, K., Toth, N. (eds). Stone Age Institute Press, Gosport, 59–82.

Hadjouis, D. (2006). Variations anatomiques posturales chez les populations franciliennes du Moyen Âge au XVIIe siècle (Val-de-Marne, France). L'exemple des dysplasies de la hanche et des malformations articulaires du genou. *Biom. Hum. Et Anthrop.*, 24(3/4), 221–232.

Harmand, S., Lewis, J.E., Feibel, C.S., Lepre, C.J., Prat, S., Lenoble, A., Boes, X., Quinn, R.L., Brenet, M., Arroyo, A. et al. (2015). 3.3-million-year-old stone tools from Lomekwi 3, West Turkana, Kenya. *Nature*, 521(7552), 310–315. doi: 10.1038/nature14464.

Heim, J.L. (1986). Compte rendu. Anatomie et évolution des veines méningées chez les Hommes fossiles. *Bull. Mém. Soc. Anthropol.*, Paris, 3(1), 60–62.

Hui, J. and Balzeau, A. (2023a). Investigating the relationship between cranial bone thickness and diploic channels: A first comparison between fossil *Homo sapiens* and *Homo neanderthalensis*. *Anat. Rec.* doi: 10.1002/ar.25360.

Hui, J. and Balzeau, A. (2023b). The diploic venous system in *Homo neanderthalensis* and fossil *Homo sapiens*: A study using high-resolution computed tomography. *Am. J. Biol. Anthropol.*, 182(3), 412–427. doi: 10.1002/ajpa.24843.

Jacobs, H., Van Boxtel, M., Heinecke, A., Gronenschild, E., Backes, W.L., Ramakers, I., Jolles J., Verhey, F. (2012). Functional integration of parietal lobe activity in early Alzheimer disease. *Neurology*, 78, 352–360. doi: 10.1212/WNL.0b013e318245287d.

Jensen, B., Agger, P., de Boer, B.A., Oostra, R.-J., Pedersen, M., van der Wal, A.C., Nils Planken, R., Moorman, A.F.M. (2016). The hypertrabeculated (noncompacted) left ventricle is different from the ventricle of embryos and ectothermic vertebrates. *Biochimica et Biophysica Acta: Molecular Cell Research*, 1863(7), 1696–1706. doi: 10.1016/j.bbamcr.2015.10.018.

Labra, N., Mounier, A., Leprince, Y., Rivière, D., Didier, M., Bardinet, E., Santin, M.D., Mangin, J.F., Filippo, A., Albessard-Ball, L. et al. (2023). What do brain endocasts tell us? A comparative analysis of the accuracy of sulcal identification by experts and perspectives in palaeoanthropology. *J. Anat.* doi: 10.1111/JOA.13966.

Leonard, W.R. and Robertson, M.L. (1997). Comparative primate energetics and homind evolution. *Am. J. Phys. Anthropol.*, 102, 268–281.

Leonard, W.R., Snodgrass, J.J., Robertson, M.L. (2007). Effects of brain evolution on human nutrition and metabolism. *Review. Annu. Rev. Nutr.*, 27, 311–327. doi: 10.1146/annurev.nutr.27.061406.093659.

Pélissier-Hermitte, B. (2006). Foramens émissaires et évolution humaine. *Antropo*, 11, 209–221.

Píšová, H., Rangel-de Lazaro, G., Veleminsky, P., Bruner, E. (2017). Craniovascular traits in anthropology and evolution: From bones to vessels. *J. Anthropol. Sci.*, 95. doi: 10.4436/JASS.95003.

Prat, S. (2024). Cognitive abilities and identity of Early Pleistocene (2.6-0.8 million years) stone tool makers, an overview. *Lesedi: Field notes*, 26, 38–42.

Rangel-de Lazaro, G., De La Cuétara, J.M., Píšová, H., Lorenzo, C., Brune, E. (2016). Diploic vessels and computed tomography: Segmentation and comparison in modern humans and fossil hominids. *Am. J. Phys. Anthropol.*, 159(2). doi: 10.1002/ajpa.22878.

Richards, G.D., Jabbour, R.S., Guipert, G., Defleur, A. (2022). Endocranial anatomy of the Guercy 1 early Neanderthal from Baume Moula-Guercy (Soyons, Ardèche, France). *Anat. Rec.* doi: 10.1002/ar.25118.

Ríos, L., Sleeper, M.M., Danforth, M.D., Murphy, H.W., Kutinsky, I., Rosas, A., Bastir, M., Gómez-Cambronero, J., Sanjurjo, R., Campens, L. et al. (2023). The aorta in humans and African great apes, and cardiac output and metabolic levels in human evolution. *Sci. Rep.*, 13, 6841. doi: 10.1038/s41598-023-33675-1.

Rival, T., Mayeur, N., Minville, V., Fourcade, O. (2009). Practical means of temperature control. *Ann. Fr. Anesth. Reanim.*, 28, 358–364. doi: 10.1016/j.annfar.2009.02.024.

Saban, R. (1982). Les empreintes du réseau vasculaire duremérien du frontal de l'homme d'Hahnôfersand, d'après le moulage endocrânien. *Bull. Mém. Soc. Anthropol.*, Paris, 9(4), 309–320.

Saban, R. (1984). Anatomie et évolution des veines méningées chez les Hommes fossiles. Comité des travaux historiques et scientifiques, Mémoire de la section des sciences, Paris.

Saban, R. (1986). L'ordre des primates : systématique et anatomie comparée. In *L'Homme, son évolution, sa diversité. Manuel d'anthropologie physique*, Ferembach, D., Susanne, C., Chamla, M.C. (eds). Éditions CNRS/Doin, Paris, 73–114.

Saban, R. (1987). Les vaisseaux méningés chez l'homme fossile de Spy, d'après les moulages endocrâniens. *Bull. Soc. roy. belge Anthrop. Préhist.*, 98, 159–174.

Saban, R. (1993). *Aux sources du langage articulé.* Masson, Paris.

Saban, R. (2007). Le sinus de Breschet. *Histoire des sciences médicales*, 11(1), 113–116.

Sorrentino, R., Stephens, N.B., Marchi, D., DeMars, L.J., Figus, C., Bortolini, E., Badino, F., Saers, J.P., Bettuzzi, M., Boschin, F. et al. (2021). Unique foot posture in Neanderthals reflects their body mass and high mechanical stress. *J. Hum. Evol.*, 161. doi: 10.1016/j.jhevol.2021.103093.

Squire, L.R. (2012). *Fundamental Neuroscience.* Elsevier, Amsterdam/Boston.

Still, A.T. (2012). *Ostéopathie : recherche et pratique.* Sully, Paris [Original published 1910].

Tournebize, R. and Chikhi, L. (2024). Ignoring population structure in hominin evolutionary models can lead to the inference of spurious admixture events. *Nat. Eco. Evo.* doi: 10.1038/s41559-024-02591-6.

Wang, H., Kim, M.N., Kieran, P., Llano, D. (2016). Thermal regulation of the brain – An anatomical and physiological review for clinical neuroscientists. *Front. Neurosci.*, 9, 528. doi: 10.3389/fnins.2015.00528.

Weaver, T.D., Coqueugniot, H., Golovanova, L.V., Doronichev, V.B., Maureille, B., Hublin, J.-J. (2016). Neonatal postcrania from Mezmaiskaya, Russia, and Le Moustier, France, and the development of Neandertal body form. *Proc. Nat. Acad. Sci. USA*, 113, 6472–6477. doi: 10.1073/pnas.1523677113.

3

Virtual Reconstruction of the Skull Base of *Sinanthropus* (*Homo erectus pekinensis*)

3.1. Introduction

Sinanthropus pekinensis is an extinct hominin, whose name is no longer used, although it allows paleoanthropologists and prehistorians to situate it in northeastern China. The fossils are classified at present in the genus *Homo* and the species *erectus*. The first skull was extracted in 1929 from Zhoukoudian cave, located 50 km southwest of Beijing, by geologist Pei Wenzhong, in the deepest layers of locus E (Jia 1999) and dated 770 ± 0.08 ka (Shen et al. 2009). The hominid lived in the very beginning of the Middle Pleistocene, which began at 774 ka, and corresponds to a glacial cooling of the northern hemisphere. Fossils of hominid populations living in Eurasia approximately 800 ka are rare, but the archeological remains are sufficient to attest to their presence from England (footprints in Happisburgh, Norfolk), to the Sunda Islands (Indonesia, Flores) via Spain (mastering fire, Cueva Negra), Israel (traces of cooking in Gesher Benot Yaaqov, north of the Sea of Galilee) and South China (Nalai site, in the Bose Basin, Acheulean lithic industry with "bifacials").

This specimen was the first almost-complete neurocranium of a fossil species discovered in mainland Asia. It is referenced as "skull III". Doctor Davidson Black (1931) undertook its comprehensive study. The suborbital torus is well formed and attests to an advanced stage of facial growth, but the cranial sutures were not synostosed; therefore, the specimen is considered a subadult with a cranial capacity of 964.4 ± 27 cc (Black 1933). The base, however, is incomplete, missing parts of the ethmoid, the body of the sphenoid, the basi-occipital, the exo-occipitals and

---

Chapter written by Tan-Nhu NGUYEN, Anne DAMBRICOURT MALASSÉ, Marie-Christine HO BA THO, Fabienne LALLOUET and Tien-Tuan DAO.

fragments of the occipital squama of cartilaginous origin. Black was able to disarticulate the seven bones and cast each one (see Figure 3.1). The surface of the endocranium was also replicated, and the missing areas were filled in with plaster, constrained by the contours of the preserved morphology. Black proposed a less inclined brain-stem orientation than in *Homo sapiens*, consistent with the higher position of the two cerebellar fossae, dictating the position of the foramen magnum (the greater foramen magnum) and thus the orientation of the cervical spine. The other specimens, discovered later in shallower layers, are adults; their bases were not preserved, and only the outer face of the skulls was cast. A total of 14 skulls, 11 mandible fragments, 147 teeth and very rare bones of the post-cephalic skeleton (only 11 bones, whereas one skull has 184 corresponding post-cranial bones) constituted the best-documented collection for *Homo erectus* populations from mainland Asia, having lived at the same site for more than 400,000 years. The fossils disappeared during the Second Sino-Japanese War (1937–1945) in 1941, when they were planned to be evacuated to the American Museum of Natural History in New York.

Skull III is therefore the only one whose endocranial surface is still visible thanks to the castings. There are only three examples existing in the world of separately cast bones: in Paris, at the Institut de paléontologie humaine (Institute of Human Paleontology) (IPH) since 1932, thanks to Henri Breuil (who directed Pei Wenzhong's thesis); in London, at University College London (UCL) since Grafton Elliot Smith's visit to Zhoukoudian in 1930 (Manias 2015); and in New York, at the American Museum of Natural History, thanks to Franz Weidenreich, the successor of Black, who had taken them with him when he was forced to leave Beijing in 1941 (Gregory 1949).

When this fossil was discovered, paleoanthropologists had very little information regarding the posterior base of the skull, even though it is, in a way, the guiding thread of the hominization process, characterized by the straightening of the axial skeleton from the basi-sphenoid to the coccyx. This straightening can be measured by angles such as the sphenoidal angle (see Figure 3.2). The closure of this angle is one of the first observations made since the comparative anatomy of modern and fossil primates. The most primitive, prosimians, have, like other mammals, a flat skull base (angle close to 180°). Current great apes, such as the chimpanzee, gorilla and orangutan, have a semi-straightened posterior base with the same sphenoid angle at birth, inherited from fossil great apes. *Homo sapiens* has the most vertically oriented posterior base. The process of straightening, since the extinct species of great ape, was still misunderstood in the 1980s. It is an embryonic process triggered by the internal dynamics of the neural tube, long before the formation of the cerebral hemispheres (Dambricourt Malassé 1988, 1993, 2002, 2006, 2011, 2022; Dambricourt and Lallouet 2009; Chaline and Marchand 2010; Dambricourt Malassé et al. 2025, and the Introduction to the current book).

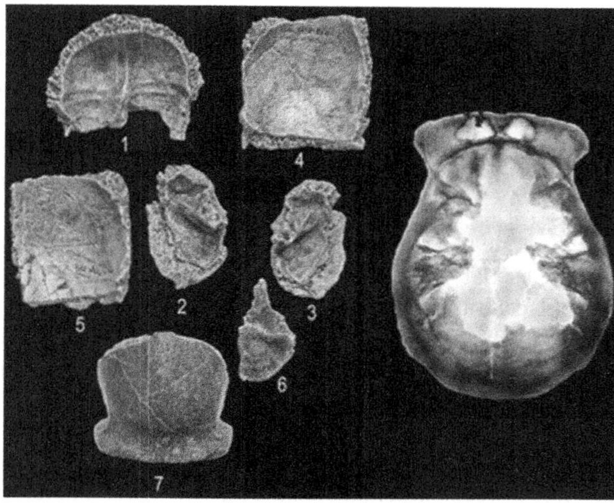

Figure 3.1. Sinanthropus *III: Davidson Black's 7 castings from 1932, preserved at the Institut de paléontologie humaine in Paris, and the radiography of the original skull (Black 1931). 1: occipital, 2 and 3: temporal squama, 4 and 5: parietal, 6: lower left wing of the sphenoid, 7: frontal bone (photo and copyright: A. Dambricourt Malassé). For a color version of this figure see www.iste.co.uk/dambricourtmalasse/futureofsapiens.zip*

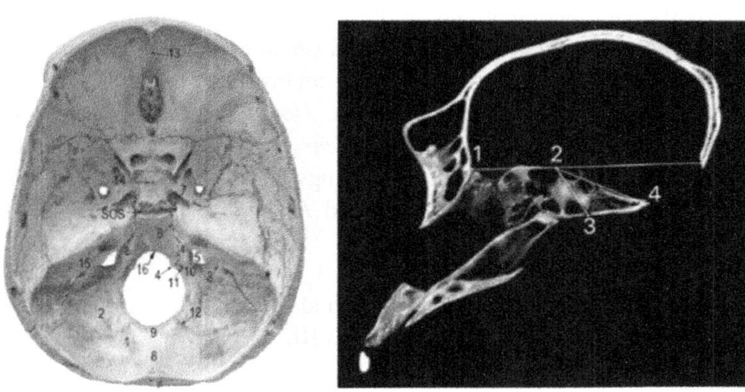

a) b)

Figure 3.2. *Base of the skull. For a color version of this figure see www.iste.co.uk/dambricourtmalasse/futureofsapiens.zip*

COMMENT ON FIGURE 3.2.– *a)* Norma verticalis *of the internal base of an adult skull of modern* Homo sapiens. *SOS: spheno-occipital synchondrosis. Homologous points for measuring the sphenoidal angle: 13: foramen caecum, 14: ephippion, 16: basion. b) CT scan of an adult chimpanzee skull in median sagittal section. Homologous points: 1: foramen caecum, 2: ephippion, 3: trace of the spheno-occipital synchondrosis on the pharyngeal face of the clivus, 4: basion. The sphenoidal angle is measured at the vertex of 1-2-4 (collections of the Institut de paléontologie humaine in Dambricourt Malassé et al. (2025)). Photos and copyright Dambricourt Malassé.*

It has therefore become necessary to study the largest number of extinct hominid skulls to compare their axial straightening angle to better understand the modalities of the evolution of the central nervous system according to the different lineages that have disappeared, and that at the origin of our species, *Sapiens*. But the majority of fossils require either a restoration following post-mortem deformations or a reconstruction of the missing or very incomplete base.

The adolescent skull of *Sinanthropus* III is the first fossil hominin whose base has been tested for a reconstruction with new MRI (computed tomography) technologies to measure the straightening angles of the posterior base, such as the sphenoidal angle (Dambricourt Malassé et al. 2025).

The castings preserved at the Institut de paléontologie humaine in Paris allowed a first reconstruction of the missing internal surface, carried out at the University of Technology of Compiègne (UTC) by means of their surface scanning followed by their virtual assembly. The 3D printing of the model made it possible to complete the missing parts with modeling clay at the Institut de paléontologie humaine. The prototype was then scanned at the American Hospital of Paris (AHP) and compared to CT scans (computed tomographs) of 55 anonymized patients, the skulls of 41 chimpanzees of growing age, 3 Javanese *Homo erectus*, 5 European *Homo neanderthalensis* and 2 European *Homo sapiens* from the Upper Pleistocene. The results distinguish three increasing straightenings: 1) the chimpanzee (Pan); 2) the extinct species of *Homo* with *H. erectus* and *H. neanderthalensis* and 3) *Homo sapiens* (Dambricourt Malassé et al. 2025).

A new study was carried out using a method to virtually complete the missing parts of the three cranial fossae of *Sinanthropus* III.

3.2. Materials and methods

3.2.1. *Materials*

The material is composed of CT scans produced at the AHP: a chimpanzee (Pan), a current *Homo sapiens* and the skull of *Sinanthropus* III, raised with plaster castings (see Figure 3.3), and that of a fossil hominin from Java, Ngandong 12, produced at the Centre hospitalier national d'ophtalmologie des Quinze-Vingts (Quinze-Vingts national ophthalmology hospital) in Paris (see Figure 3.4). *Sinanthropus* and Ngandong hominins belong to the species *Homo erectus*, but Ngandong is among the last representatives of the Indonesian species, dying out approximately 117–108 ka (Rizal et al. 2020). This skull is not complete; it lacks part of the upper fossa of the base with the greater wings of the sphenoid and the anterior edge of the sella turcica (see Figure 3.4(d)), and its vault is higher because the brain was more developed. The inclination of the cerebellar face of the petrous pyramids is identical, however, and this is what matters for positioning the inclination of the clivus.

The method developed makes it possible to virtually fill in the missing parts of the three basal fossae, measure the sphenoidal angle and test the validity of the reconstructed position of the clivus that supports the brain stem in adolescent *Sinanthropus* III. The method proceeds in several steps.

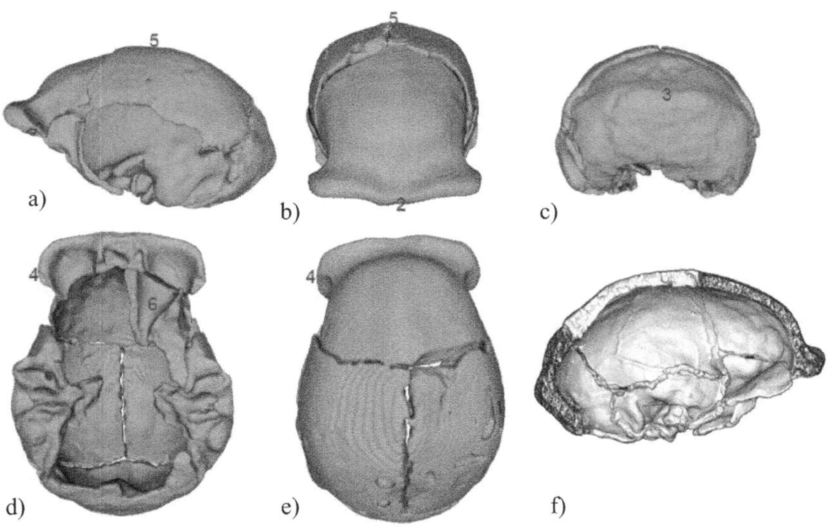

Figure 3.3. *CT scan of the plaster replica of the skull of* Sinanthropus *III, recomposed with the assembly of the isolated castings. The recesses visible on the parietals correspond to the bubbles formed in the plaster. For a color version of this figure see www.iste.co.uk/dambricourtmalasse/futureofsapiens.zip*

COMMENT ON FIGURE 3.3.– *a) Left-external* norma lateralis, *b) plunging view on the frontal bone with the supraorbital torus, c)* norma occipitalis, *d)* norma basilis, *e)* norma verticalis, *f) median sagittal section internal view of the left hemicranium. 1: porion, 2: glabella, 3: inion, 4: most-lateral edge of the supraorbital arch, 5: vertex, 6: slightly-offset fragment of the greater wing of the sphenoid.*

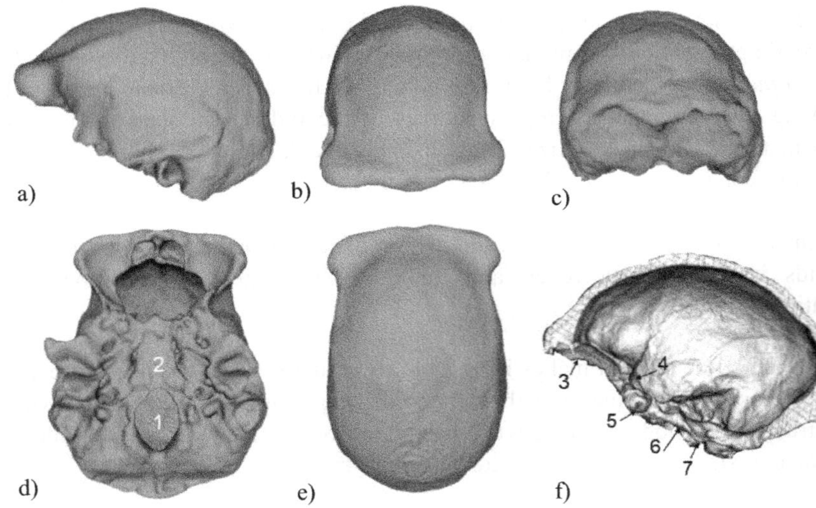

Figure 3.4. *CT scan of Ngandong 12. For a color version of this figure see www.iste.co.uk/dambricourtmalasse/futureofsapiens.zip*

COMMENT ON FIGURE 3.4.– *a) Left* norma lateralis, *b) plunging view on the frontal bone with the supraorbital torus, c)* norma occipitalis, *d)* norma basilis, *e)* norma verticalis, *f) median sagittal section in the internal view of the right hemicranium, the body of the sphenoid is incomplete with the disappearance of the sella turcica. 1: foramen magnum, 2: clivus, 3: foramen caecum, 4: approximate location of the ephippion, 5: sinus of the body of the sphenoid, 6: clivus, 7: basion.*

3.2.2. *Meshing of the virtual images*

The first step is meshing (see Figure 3.5), which consists of the geometrical modeling of a continuous medium in the form of interconnected points. The points of the numerically constituted space are called the vertices of the mesh. The vertices may be connected three-by-three, forming triangular polygonal cells, or facets (see Figure 3.6). The coordinates in the space of the vertices can be modified virtually.

The digitization meshing (or computed tomography) of *Sinanthropus* III is the target, while the CT scans of the chimpanzee, Ngandong 12 and *Homo sapiens* serve as a reference. As shown by the comparison between Figures 3.3(a) and (c) and 3.4(a) and (c), the height of the skull vault above the porion (1) and the inion (3) is significantly less in *Sinanthropus* III, but the structure seems the same as that of Ngandong 12 with an elongated shape, an occipital squama with a nuchal plane oriented backward and upward, and not forward and downward like *Sapiens*. The three skulls were also subjected to a meshing to test one that offers the best conformity to reconstruct the position and shape of the bones composing the three cranial fossa of *Sinanthropus* III.

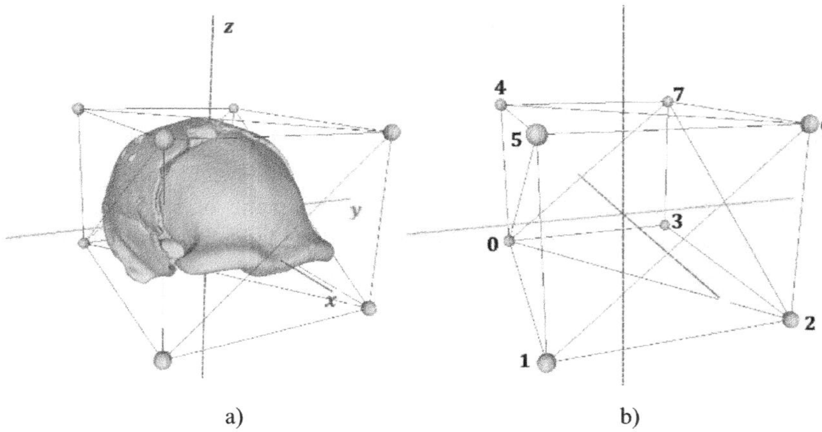

Figure 3.5. *a) Digitization of the virtual image of* Sinanthropus *III in the 3D reference point of a bounding box. b) Bounding box. For a color version of this figure see www.iste.co.uk/dambricourtmalasse/futureofsapiens.zip*

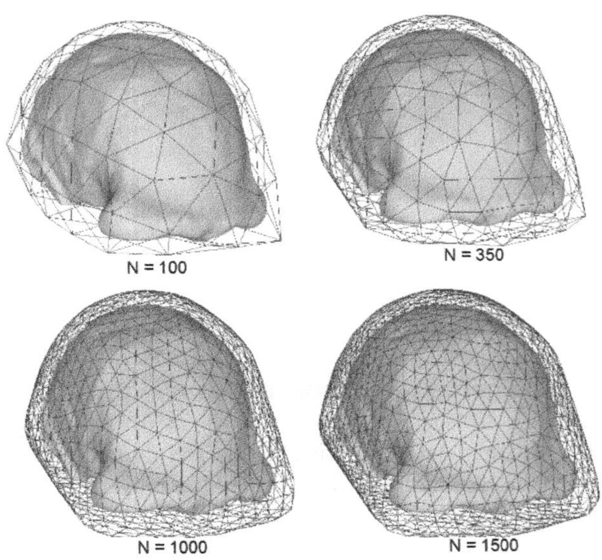

Figure 3.6. *Mesh of the Ngandong 12 digitization in a deformation control cage with different numbers of control points (N), allowing its shape to be adjusted to that of another skull*

3.3. Reconstruction of the three basioccipital fossae of *Sinanthropus* III

The second step consisted of reconstructing the three basioccipital fossae of *Sinanthropus* III.

3.3.1. *The upper fossa*

The fragment of the greater wing of the sphenoid that extends the left orbit (no. 6 in Figures 3.1 and 3.3(d)) has made it possible to fill in the right part by symmetry. The bounding box of the mesh was divided in the plane of symmetry, and the missing area was completed by reflecting the vertices and facets of the left part; then, the two parts were merged (see Figure 3.7).

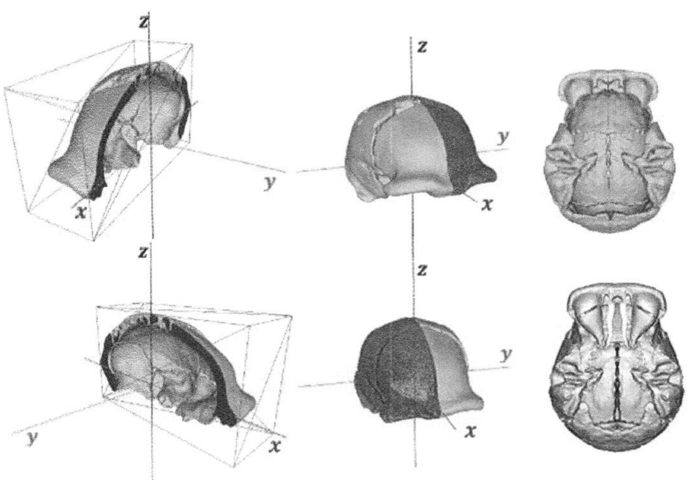

Figure 3.7. *Reconstruction of the upper cerebral fossa of* Sinanthropus *III by symmetry with the left hemicranus (bottom mesh). For a color version of this figure see www.iste.co.uk/dambricourtmalasse/futureofsapiens.zip*

3.3.2. *Middle and posterior (or cerebellar) fossae*

It was possible to complete the missing part of the middle and posterior fossae using the mesh from the Ngandong 12 digitization. We proceeded in three steps (see Figure 3.8). The first step consists of placing in the same bounding box the symmetrized digitization of *Sinanthropus* III and that of Ngandong 12, identifying the five defined homologous points: the porion, the inion on the occipital protuberance, the glabella, the most-lateral edges of the supraorbital torus and the vertex of the vault. Next, the Ngandong 12 mesh was transformed into the *Sinanthropus* mesh coordinate system, rigidly, i.e. manually, by taking the anatomical points. The transformation was performed using the singular value decomposition (SVD) method. However, the five homologous points not being sufficient to deform the surfaces, we added the iterative closest point (ICP) algorithm to the deformation control cage, allowing it to be applied to the points closest to the vertices of each of the two meshes. Step 2 consisted of repositioning the vertices of the Ngangdong 12 mesh with the control points, adjusting them as closely as possible to those of *Sinanthropus* III, resulting in the hybrid exact-geometry model. Step 3 transforms the "hybrid" mesh – or exact-geometry model – into a convex shape or hull.

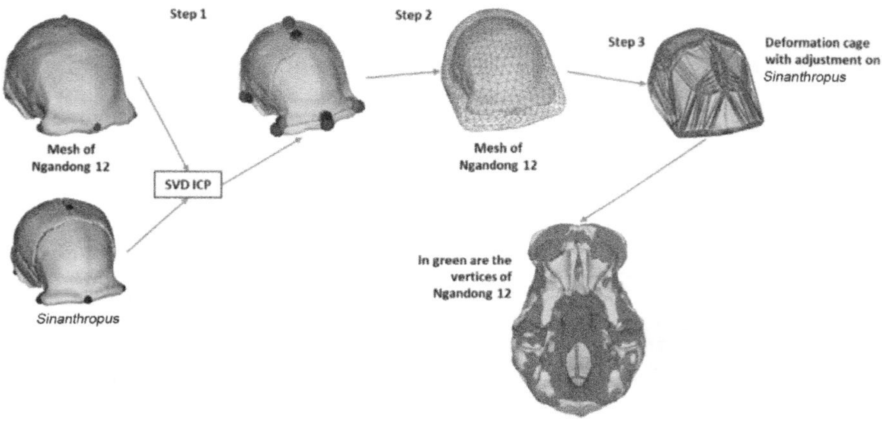

Figure 3.8. *The three stages of the Ngandong 12 superposition, adjusted to* Sinanthropus *III, up to the hybrid model. Step 3 transforms the hybrid mesh into a convex hull. Explanations are in the text. For a color version of this figure see www.iste.co.uk/dambricourtmalasse/futureofsapiens.zip*

The middle and posterior fossae of *Sinanthropus* III are reconstructed with the body of the sphenoid and the clivus of Ngandong 12, hybridized. All of the green vertices of the base were manually cropped (on screen) and adjusted acceptably to the *Sinanthropus* III mesh; then, the two were merged (see Figure 3.9).

The skull base is formed of orifices for the passage of nerves and blood vessels, as well as sutures (synchondroses). Present on Ngandong 12, these spaces return to their original location on *Sinanthropus* III. The inclination of the clivus and the position of the basion are restored and measurable. The virtual reconstruction can be compared to that reconstructed with modeling clay (see Figure 3.10).

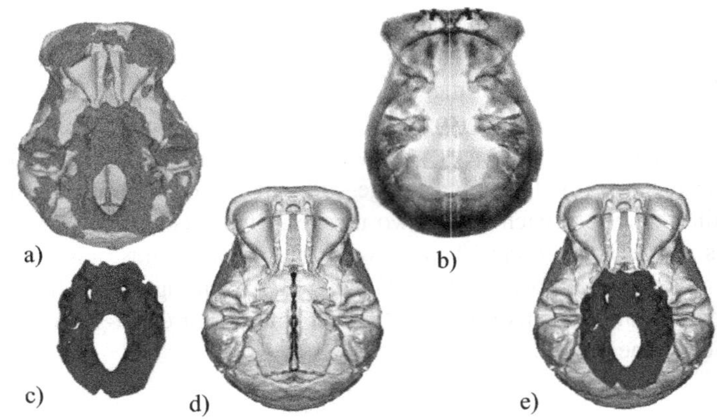

Figure 3.9. Norma basalis*: a) hybrid Ngandong 12; b) digital X-ray image of the original skull of* Sinanthropus *III symmetrized (on screen); c) middle and posterior base of the hybrid skull, manually cropped (on screen); d) symmetrized* Sinanthropus *III; and (e) virtual reconstruction of the middle and posterior base by merging. For a color version of this figure see www.iste.co.uk/dambricourtmalasse/futureofsapiens.zip*

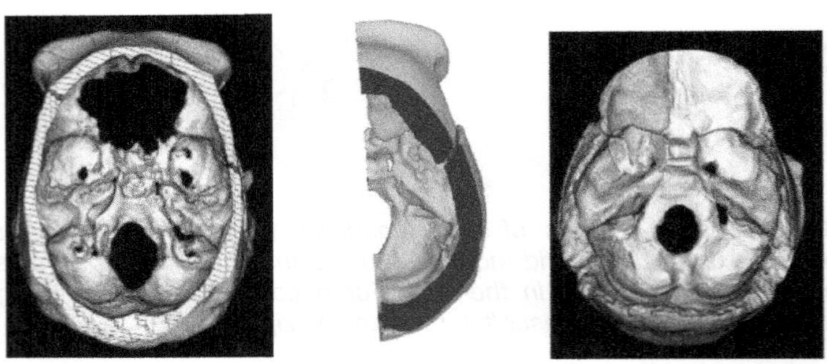

a) b) c)

Figure 3.10. *CT scans in* norma verticalis *of the internal base: a) Ngandong 12; b)* Sinanthropus *III reconstructed virtually; c)* Sinanthropus *III reconstructed manually (in white) (A and C in Dambricourt Malassé et al. (2025)). For a color version of this figure see www.iste.co.uk/dambricourtmalasse/futureofsapiens.zip*

3.4. Validation

A validation procedure was conducted to select the best cranial digitization with the Hausdorff distance (see Figure 3.11).

This method compares the *Sinanthropus* III target mesh and the meshes of the chimpanzee, the hybrid model of Ngandong 12 and *Homo sapiens*. For each mesh type, we increased the number of control points from 50 to 2,800 points, with a step size of 50 points. The optimal distance errors from one mesh to another were collected for evaluation. The extremes (best and worst validations) were analyzed in detail, using color distance maps (see Figure 3.12). The scale is inframillimetric.

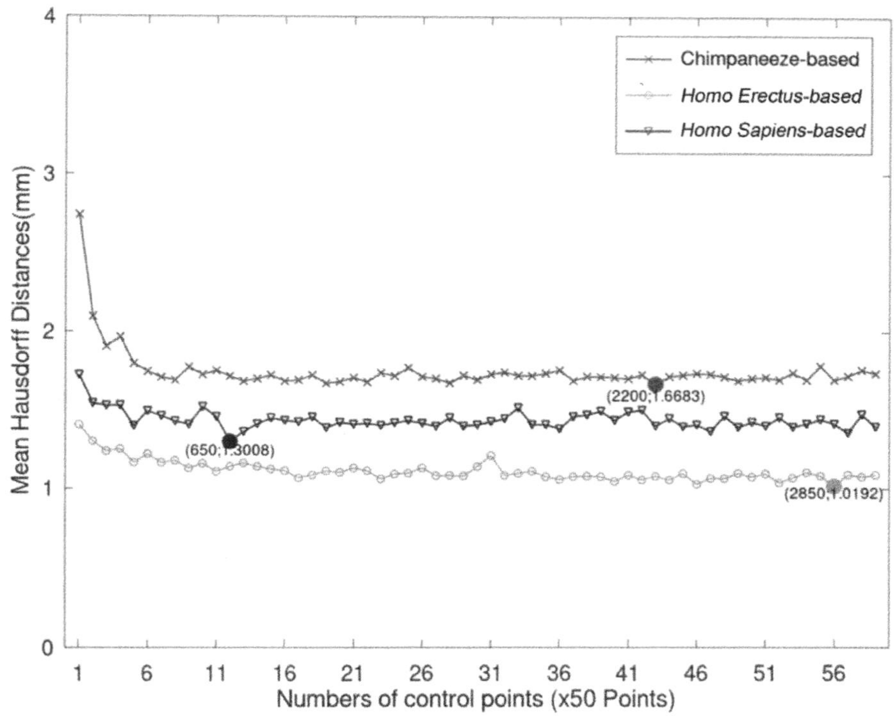

Figure 3.11. *Mean Hausdorff distances between the* Sinanthropus *target skull mesh and the meshes of the chimpanzee, the Ngandong 12 hybrid model, and* Homo sapiens*, as a function of the number of control points*

When the chimpanzee mesh is selected with 2,200 optimal control points, the mean adjustment error is 1.66 mm. When the mesh is *Homo sapiens*, the mean errors are 1.30 mm with a lower number of control points, i.e. 650 points. The hybrid Ngandong 12 mesh with the optimal number of 2,850 control points is the one that presents the minutest mean adjustment error, at 1.01 mm.

As predicted, the hybrid skull generated from the Ngandong 12 mesh shows fewer errors than that of the chimpanzee and *Homo sapiens*. The best- and least-adapted cases for the shapes (i.e. the convex hull of the skull) and the meshes or models of the skull (i.e. the exact geometry) are shown in Figure 3.12. In the worst cases, the adjustment errors for the shapes (mean ± standard deviation) are 1.41 ± 1.50 mm, and those for the skull models are 2.43 ± 3.15 mm. In the best-predicted case, the adjustment errors are 1.02 ± 1.40 mm and 2.21 ± 3.19 mm for the shapes and models, respectively. We note that in comparison with the vault shapes, the adjustment distances most exposed to errors (in blue) are those of the middle and posterior base: position of the body of the sphenoid and the basion, which is not surprising since this part is incomplete on Ngandong 12.

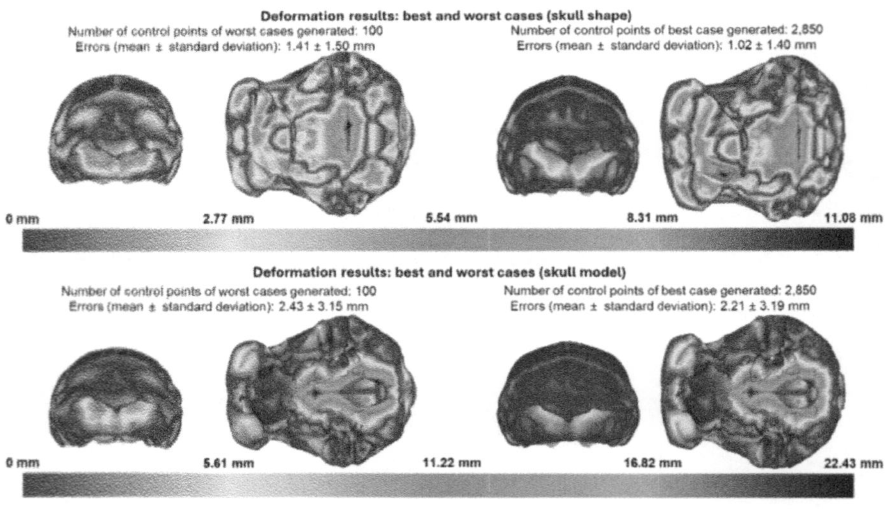

Figure 3.12. *The best and worst cases were predicted when the number of control points was 100 and 2,850, at the top for the shape (i.e. the convex hull), and at the bottom for the model (i.e. the exact geometry) of the skull. For a color version of this figure see www.iste.co.uk/dambricourtmalasse/futureofsapiens.zip*

3.5. The sphenoid angle: discussion

The mean value of the sphenoidal angle measured on the series of adolescent and adult chimpanzees is 162° and 165°, respectively. It is 125° and 127° in modern *Homo sapiens*. The angle of the plaster reconstruction of *Sinanthropus* III is 135°. The body of the sphenoid of Ngandong 12 is incomplete; the sella turcica is destroyed, and therefore the ephippion is missing. The foramen caecum, on the other hand, can be localized, the basion is preserved and the median axis of the clivus can be traced. The value of the sphenoidal angle of Ngandong 12 is approximated and estimated at 138°; that of two other Javanese *Homo erectus* is 138° for Ngandong 7 and 129° for Ngawi (Dambricourt Malassé et al. 2025). The approximate sphenoidal angle of the virtual reconstruction of *Sinanthropus* III is 137°.

3.6. Conclusion

The techniques of mesh digitization and deformation can be applied to the reconstruction of incomplete or deformed fossilized bone structures from complete, current and/or fossil bone structures taken as reference. For the cranial base of *Sinanthropus* III, an adolescent *Homo erectus pekinensis* of 700 ka, we chose the SVD method, supplemented by the ICP algorithm, applied to the deformation control cage. Three reference skulls, that of a chimpanzee, a *Homo erectus* from Java that is geologically more recent (Ngandong 12) (110–100 ka) and a *Homo sapiens,* were able to confirm the difference in angular conformation, in the sense of an intermediate verticalization between the great ape and contemporary human anatomy. This degree of straightening remained unchanged for 600,000 years within the Asian *Homo erectus* lineage, while the cranial vault rose and lengthened with the complexification of the different lobes of the cerebral hemispheres. The sphenoidal angle of *Sinanthropus* III remains approximated, but the method shows that it will always be possible to better specify it thanks to comparisons with better-preserved *Homo erectus* skulls such as Ngandong 7, or future fossils to be discovered.

3.7. Acknowledgments

This study formed part of the theme "Hominids in the Quaternary: Environments and Behaviors" (HQEB), which was led by François Sémah and Marie-Hélène Moncel of UMR 7194 CNRS – Muséum national d'histoire naturelle. We are grateful to François Sémah for having made available to us the CT scan of Ngandong 12 produced in Paris, at the Centre hospitalier national d'ophtalmologie des Quinze-Vingts.

3.8. References

Black, D. (1931). On an adolescent skull of *Sinanthropus pekinensis* in comparison with an adult skull of the same species and with other hominid skulls, recent and fossil. *Palaeontologia Sinica D*, 7(2), 1–144.

Black, D. (1933). On the endocranial cast of the adolescent Sinanthropus skull. *Proceedings of the Royal Society of London. Series B*, 112, 263–276.

Chaline, J. and Marchand, D. (2010). *Le singe, l'embryon et l'homme. Une nouvelle clé de lecture de l'histoire de l'homme*. Ellipses, Paris.

Dambricourt Malassé, A. (1988). Hominisation et fœtalisation. *Comptes Rendus de l'Académie des Sciences*, 307(II), 199–204.

Dambricourt Malassé, A. (1993). Continuity and discontinuity during hominization. *Quaternary International*, 19, 85–98. doi: 10.1016/1040-6182(93)90027-D.

Dambricourt Malassé, A. (2002). Nouvelles approches de l'évolution crânienne des *Homo erectus* de Java. In *Origine des peuplements et chronologie des cultures paléolithiques dans le Sud-Est asiatique*, Sémah, F., Falguères, C., Grimaud Hervé, D., Sémah, A.M. (eds). Artcom', Paris, 377–396.

Dambricourt Malassé, A. (2006). Évolution du chondrocrâne et de la face des grands anthropoïdes miocènes jusqu'à *Homo sapiens*, continuités et discontinuités. *Comptes Rendus Palevol*, 5(1–2), 109–117. doi: 10.1016/j.crpv.2005.09.024.

Dambricourt Malassé, A. (2011). Les équilibres bipèdes permanents, origine embryonnaire, morphogenèse, équilibre occluso-postural, conséquences pour l'évolution psychomotrice et comportementale des hominidés. HDR Thesis, Université de Technologie de Compiègne, Compiègne.

Dambricourt Malassé, A. (2022). *Embryogeny and Phylogeny of the Human Posture 2. A New Glance at the Future of Our Species*. ISTE Ltd, London.

Dambricourt Malassé, A. and Lallouet, F. (2009). Étude au scanner des canaux semi-circulaires d'un crâne sec d'anencéphale dépourvu de flexion sphéno-occipitale ; implications pour l'interprétation paléo-embryologique des hominidés fossiles. *Biométrie humaine et anthropologie*, 27(1/2), 79–86.

Dambricourt Malassé, A., Ho Ba Tho, M.-C., Dao, T.T., Lallouet, F. (2025) Reconstruction of the internal cranial base of an adolescent *Homo erectus* (*Sinanthropus* III). An enlightening on the hominids neural straightening and its cognitive consequences. *L'Anthropologie*, 3.

Gregory, W.K. (1949). Franz Weidenreich (1873–1948). *American Anthropologist*, 51, 85–90.

Jia, L. (1999). *Chronicle of Zhoukoudian (1927–1937)*. Shangai Scientific and Technical Publishers, Beijing.

Manias, C. (2015). Sinanthropus in Britain: International science and the nature of humanity, 1920–1939. *British Journal for the History of Science*, 48(2), 289–319. doi: 10.1017/S0007087414000909.

Rizal, Y., Westaway, K.E., Zaim, Y., van den Bergh, G.D., Bettis, E.A. III., Morwood, M.J., Huffman, O.F., Grun, R., Joannes-Boyau, R., Bailey, R.M. et al. (2020). Last appearance of *Homo erectus* at Ngandong, Java, 117,000-108,000 years ago. *Nature*, 577, 381–385. doi: 10.1038/s41586-019-1863-2.

Shen, G., Gao, X., Gao, B., Granger, D.E. (2009). Age of Zhoukoudian *Homo erectus* determined with 26Al/10Be burial dating. *Nature*, 458(7235), 198–200. doi: 10.1038/nature0774.

4

Characterization and Modeling of Morphological Evolutions in Bones

4.1. Introduction

Numerical methods are appropriate for modeling bones and joints. These tools are well-suited to modeling and taking into account the geometric complexities as well as the physical constituents that compose them. Bones are complex structures and systems, both morphologically and mechanically, but they also evolve over time, as they are living structures. Indeed, there are several levels of organization; bone is a composite, anisotropic and heterogeneous material. In addition, inter-subject variations exist, and are to be taken into consideration as necessary.

Biomechanics, which consists of understanding the mechanics of life, is appropriate for studying the natural morphological evolution of bones, from growth to aging. Bone-related pathologies generally demonstrate a malformation, a deformation sometimes associated with an alteration in the mechanical properties. To take into account these differences, we have developed a modeling methodology that simultaneously integrates the geometric complexity and mechanical characteristics of the bone and osteoarticular system via medical imaging techniques. The methodology is based on the semi-automatic generation of anatomical contours and surfaces of bones and joints, derived from medical image processing techniques (Ho Ba Tho 1993). Subsequently, predictive relationships combining physical properties determined via imaging techniques, and mechanical properties (Rho et al. 1995) via destructive and non-destructive (ultrasonic) techniques, will allow us to develop models with individualized geometric and mechanical properties.

Chapter written by Marie-Christine HO BA THO.

4.2. Methodology

4.2.1. *Geometric reconstruction*

There are several medical imaging modalities for examining bones and joints: standard radiographies, computed tomography (CT) images and magnetic resonance imaging (MRI). The first two techniques, based on X-ray absorption, are often used to visualize hard biological materials such as bone tissue. MRI, on the other hand, is based on hydrogen–proton resonance, favoring the visualization of soft biological materials such as ligaments, cartilage and muscles.

A methodology has been designed and developed to preprocess CT and MRI images, for geometric reconstruction dedicated to structural calculations (Ho Ba Tho 1993). The various steps consist of decoding the images, performing image processing and constructing geometric entities defined in a neutral format, or IGES. The standard coding format for medical images is called DICOM (Digital Imaging Communication in Medicine), derived from ACRNEMA (American College of Radiology and National Electric Manufacturers Association) specifications. The images consist of matrices of pixels, each pixel being coded on 12 bits (the pixel values varying from 0 (white) to 4,090 (black)). Once these images have been decoded, a thresholding based on the histogram representing the distribution of pixel values allows the binarized images to be visualized in black and white. The region-based segmentation method is used, with contour detection enabling the contours of the anatomical structure to be delineated. The interpolation curves used are Hermite curves. The different steps of geometric reconstruction of a tibia in vivo are shown in Figure 4.1. The Patran pre- and post-processors (MSC Software) are used to visualize geometric entities written in a file in ASCII format. In order to develop realistic models, finite elements will be assigned the mechanical properties associated with the anatomical structure studied.

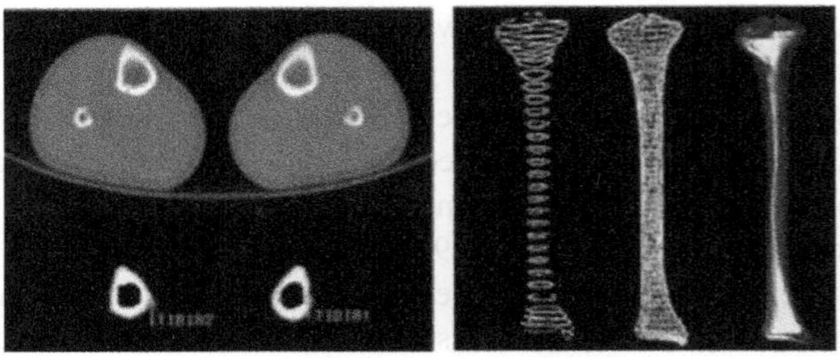

a) b)

Figure 4.1. *Contour detection of CT images (a) and three-dimensional model of the internal and external contours and surfaces of the long bone (b). For a color version of this figure see www.iste.co.uk/dambricourtmalasse/futureofsapiens.zip*

4.2.2. *Mechanical properties of human bone tissue*

The mechanical properties of bone have been studied for over three decades to understand the mechanism of bone fracture, healing and remodeling, and associated bone pathologies. Despite this, the data remained insufficient, notably for numerical modeling. Many simplifying assumptions have been widely used. Human bone is a highly heterogeneous biological material, with numerous structure levels that are more or less hierarchical. It is a composite material made up of two types of bone tissue: cortical (compact) bone and cancellous (highly porous) and biphasic bone (organic and inorganic constituents). In our characterization studies, we considered anisotropic behavior of bone tissue. In order to determine the elastic properties, we developed an ultrasonic transmission device dedicated to bone tissue (Ho Ba Tho et al. 1992). Given the different bone structures, cortical tissue is considered an unlimited medium and cancellous bone limited. The propagation velocities of the respective media were then measured. The matrix of elastic coefficients could be calculated, then by inverting it, the elastic properties were obtained. An atlas of mechanical properties of different types of cortical and cancellous human bone was obtained (femur, tibia, mandible, humerus, patella, lumbar vertebrae, etc.) (Ho Ba Tho et al. 1992, 1998a). The results showed a variation in mechanical properties, depending on the anatomical location, the bone type and the individual. The results have led us to consider these variations by developing models with individualized geometric and mechanical properties.

4.2.3. *Models with individualized geometric and mechanical properties*

The physical measurements performed on medical images will allow us to combine geometric and mechanical properties simultaneously. Indeed, we have established empirical relationships between CT measurements and mechanical property measurements performed on the same samples. The CT scan number, expressed in Houndsfield Units (HU), is a physical characteristic of CT measurements of biological tissue. The pixel values will be expressed by the empirical CT number. This number depends on the acquisition parameters and, more particularly, on the linear coefficients of X-ray attenuation through the tissue. Thus, by performing mechanical property measurements on bone samples simultaneously with CT scanning, predictive relationships have been established that allow correlations between the CT number and mechanical properties (Rho et al. 1995).

4.3. Clinical applications

The main applications will concern the quantification of bone malformations in children and their consequences on joints; congenital dysplasia of the hip (Roach et al. 1997), club feet (deformation of the foot at birth) (Johnston et al. 1995; Ho Ba Tho et al. 1998b), lower-limb malformations (Limbert et al. 1998; Périé and Ho Ba Tho 1998) and scoliosis (spinal deformation) (Périé et al. 2001). MRI images are generally used to study cartilage and bone malformations in children, quantitative analysis of these malformations being intended to plan and evaluate surgical or orthopedic treatment (correction with a splint or a corset). Figure 4.2 shows finite-element models with individualized in vivo geometric and mechanical properties of lumbar vertebrae (L1), for two scoliotic patients of the same age (Périé et al. 2002).

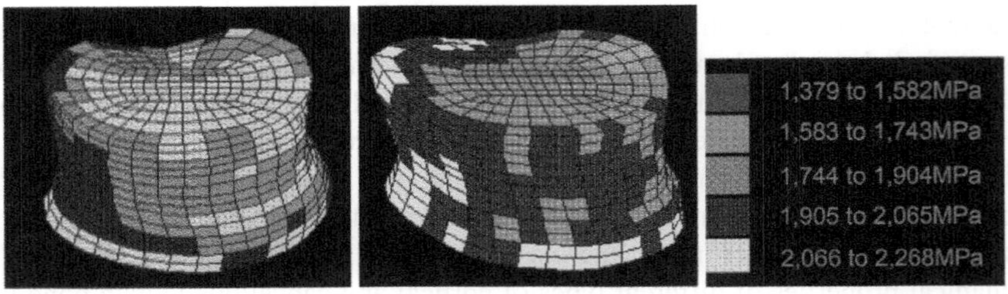

Figure 4.2. *Finite-element models of lumbar vertebrae with individualized geometric and mechanical properties of two patients with scoliosis. For a color version of this figure see www.iste.co.uk/dambricourtmalasse/futureofsapiens.zip*

4.4. Conclusion

These modeling techniques contribute to the understanding of bone malformations, and make it possible to quantify them with a view to proposing an appropriate treatment (surgery or wearing splints or corrective orthoses). Above all, our method enables a personalized evaluation, necessary for optimal preoperative surgical or rehabilitation planning. The imaging techniques used upstream of the modeling methodology are powerful. If well mastered, knowledge extraction of patients' characteristics, such as morphology, mechanical properties and data concerning load on the human musculoskeletal system, could be assessed from medical imaging (Ho Ba Tho and Dao 2019).

4.5. References

Ho Ba Tho, M.C. (1993). Logiciel SIP 305. Logiciel de prétraitement d'images médicales, Scanner, IRM.

Ho Ba Tho, M.C. and Dao, T.T. (2019). Knowledge extraction from medical imaging for advanced patient-specific musculoskeletal models. *Encyclopedia of Biomedical Engineering*, 135–142.

Ho Ba Tho, M.C., Rho, J.Y., Ashman, R.B. (1992). Atlas of mechanical properties of human cortical and cancellous bone. In *In Vivo Assessment of Bone Quality by Vibration and Wave Propagation Techniques, Part II*, Van Der Perre, G., Lowet, G., Borgwardt, A. (eds). ACCO, Leuven, 7–38.

Ho Ba Tho, M.C., Rho, J.Y., Ashman, R.B. (1998a). Anatomical variation of mechanical properties of human cancellous bone in vitro. In *Bone Research in Biomechanics*, Lowet, G., Rüegsegger, P., Weinans, H., Meunier, A. (eds). IOS Press, Amsterdam, 157–173.

Ho Ba Tho, M.C., Luu, S., Estivalèzes, E., Baunin, C., Cahuzac, J.P. (1998b). Simulation of the 3D motion of clubfoot bones using helical axes theory. *Journal of Biomechanics*, 31, 107.

Johnston, C.E., Ho Ba Tho, M.C., Baker, K.J., Baunin, C. (1995). Three dimensional analysis of clubfoot deformity using computed tomography. *Journal of Pediatric Orthopaedics*, 4(B), 39–48.

Limbert, G., Estivalèzes, E., Ho Ba Tho, M.C., Baunin, C., Cahuzac, J.P. (1998). In vivo determination of homogenised mechanical characteristics of human tibia. Application to the study of tibial torsion in vivo, *Clinical Biomechanics*, 13, 473–479.

Périé, D. and Ho Ba Tho, M.C. (1998). In vivo determination of contact areas and pressure of the femorotibial joint using non linear element analysis. *Clinical Biomechanics*, 13, 394–402.

Périé, D., Sales de Gauzy, J., Sévely, A., Ho Ba Tho, M.C. (2001). In vivo geometrical evaluation of Cheneau Toulouse Munster brace effect on scoliotic spine using MRI method. *Clinical Biomechanics*, 16, 129–137.

Périé, D., Ho Ba Tho, M.C., Baunin, C., Sales De Gauzy, J. (2002). Personalised mechanical properties of scoliotic vertebrae determined in vivo using tomodensitometry. *Computer Method in Biomechanics and Biomedical Engineering*, 5(2), 161–165.

Rho, J.Y., Ho Ba Tho, M.C., Ashman, R.B. (1995). Relations of mechanical properties to density and CT numbers in human bone. *Medical Engineering Physics*, 17(5), 347–355.

Roach, J.W., Ho Ba Tho, M.C., Baker, K.J., Ashman, R.B. (1997). Three-dimensional computer analysis of complex acetabular insufficiency. *Journal of Pediatric Orthopaedics*, 17, 158–164.

PART 2

Current Perspectives: From the Fetus to Life in Space

5

Pre- and Postnatal Haptonomy: An Early Humanization

5.1. Introduction

The point of view of haptonomy: in the baby *sapiens*, the process of humanization begins with verticalization.

In *The Human Condition*, Hannah Arendt (1983) says, "We are beings of beginnings." The particularity of humans, due to their long dependence on adult guardians, means that this capacity to inaugurate must be passed on to us. Little humans are full of treasures that they can only access if another human gives them the key. To allow the subject, an autonomous source of desire, to develop as a human, to bring them to fully deploy all of their potential, to assume their values of humanity, beyond cultural barriers, and to humanize this subject full of desires and impulses, is to "make the human a horizon", as the philosopher Marie-José Mondzain (2023) puts it.

Humans are the only mammals that are constantly bipedal, once this hard-won ability is acquired, between the age of 11 and 18 months. We know the significance of this specificity for the development of *Sapiens*, and in this book other specialists develop on its significance and consequences. We will therefore focus here on the specific contribution of early verticalization through our clinical experience of haptonomy, which we have been practicing for more than 35 years, particularly during the pre- and postnatal accompaniment of childbirth. We wish to impart the

---

Chapter written by Catherine DOLTO.

For a color version of all figures in this chapter, see www.iste.co.uk/dambricourtmalasse/futureofsapiens.zip.

understanding of why and how prenatal haptonomic contact prepares the child, from intrauterine life, to assume the verticality specific to its species. It is not a matter of making the child walk earlier, but of helping it feel, mobilize and use its spine in an encounter project before it is even born. For this demonstration to be clear, we need to describe our work in some detail.

5.2. Haptonomy

A brief presentation seems necessary here to shed light on what we discuss below. Its founder, Dutch physician and phenomenologist Frans Veldman (1921–2010), made a very simple discovery: he understood that within the psychic apparatus in its entirety, we could bring together in a consistent whole everything that concerns perceptions, sensations, emotions and feelings; this is what he named "the Affective" (Veldman 1998). His life's work was to precisely describe how it functions, and to develop a phenomenality of the human encounter, which is determined at every moment by the Affective, whereas we want to believe it is orchestrated by rationality. That is why he called haptonomy the "science of affectivity". Manifestations of the Affective are reflected in immediate changes in all tone, accompanied by dilation of the vessels and hormonal secretions. Most of our contemporaries do not pay attention to it, but we well know the spectacular physical manifestations of strong emotions that precede the appearance of feelings: flushing, pallor, accelerated heart or breathing rate, hairs standing on end, abdominal pain, etc. The Affective is governed by the subcortex, the limbic system and the subcortical pathways, via gamma fibers, which are very rapid. This is why any affective approach causes changes in tones, which become flexible and modulated, almost instantaneously in all tissues. Conversely, cortical control, transmitted by the slower alpha fibers, gives a less flexible tone, even rigidity, which obeys the law of all or nothing. At every moment, it is the balance between these two innervations that is decisive, but in Western urban dwellers, cortical control is highly dominant while subcortical control is little known.

Indeed, contemporary Western societies greatly value cognitive faculties and rationality, and educational processes are directed primarily at the left cortex, yet the key to intelligence lies in a global and synergic functioning of the two hemispheres. In prenatal life, it is the right cortex, more devoted to the Affective, that develops first. The same gesture can be controlled rationally or affectively: the experience of the person making the gesture, their feelings linked to their tones, just like the experience of the person observing it, will be different depending on whether the gesture is controlled by the cortex or the subcortex. Conduct this very simple experiment: open your arms wide then close them, several times, by rational decision: "I want to open and close my arms"; then open them as you imagine welcoming someone you love and who is in need of it. The difference is obvious, perceived from the inside, just as it can be seen from the outside. Haptonomy advocates the importance of a harmonious interaction between these two systems, without the subcortical pathways being inhibited by the usual cortical dominance.

When the Affective is awakened by an appropriate approach, an immediate change can be observed in the sense of self, via a relaxation or tensing-up at tissue and cellular level, accompanied by changes in breath, heart rate and hormonal secretions.

The Affective is thus the secret organizer of the incessant interconnecting between body and mind, allowing the human being to be approached as an affective-somatic-psychic unit, with no excluded third. This departure from dualistic thought (separate body/mind) is presented nowadays as self-evident and necessary by a great number of researchers and physicians. But for many non-professionals, these phenomena remain little known. The current success of body therapies shows just how deeply dualism remains rooted in current thinking. Only, they are expected to have psychoaffective effects that they cannot deliver, even if they are otherwise very beneficial. Haptonomy proposes a Western practice of this global approach, explainable and reproducible, without esotericism, mastered, and clear and simple to access, which constitutes undeniable progress. Although it uses a lot of tactile contact, it is not to be classified among body therapies, but as a therapy of the embodied human subject (assigned to residence in a human, carnal form) in all its dimensions.

Countless internal and external perceptions assail us every second; it is the Affective that filters those that, according to our personal history and that of our ancestors, will rise toward our consciousness or preconscious. Haptonomy provides conceptual tools: it recalls the importance of the conscious ego, an ethical entity; the non-conscious; and the preconscious. In education, haptonomy emphasizes the development of consciousness: self-awareness, then consciousness of our actions. These concepts bring very useful nuances to a line of thought that proved reductive, given that it was reduced to opposing the conscious to the psychoanalytic unconscious, a place of tensions, impulses and repressions. Haptonomy therefore constitutes a truly innovative theoretical and clinical openness. Sensory perceptions, once examined though the lens of the Affective, are assigned an indication: good or bad for me, reassuring/worrying, fair/not fair, etc. This is how the sensory becomes sensual, i.e. loaded with meaning. The Affective is no longer this mysterious and uncontrollable force, capable of overwhelming, submerging and trapping us, but a dimension of our being that we can know and use. The well-controlled implementation of the Affective in care, proves to be highly effective. It highlights and puts into action changes in tone that modify our presence in the world, leading, with a lesser or greater degree of consciousness, to an uncontrolled reaction in the person approached[1], in a reciprocity that is always present, but rarely conscious.

Haptonomy thus offers a new interpretation of the concept of intelligence, child development and psychopathology in general. It brings an entirely new reading of what is usually placed within the vague framework of psychosomatics. It uses very precise techniques in the handling of tactile contact and the construction of presence with the other, beyond ourselves, through the use of space. Haptonomy can thus be viewed as a mastered art of human encounter, which allows us to revisit all the professions of care, education and transmission. Without opposing modern medicine, it enhances it.

Frans Veldman's thinking was about 60 years ahead of Western thought. Everything he has put forward is now well informed by the discoveries of neuroscience. He emphasized:

– the paramount importance of the Affective, which constantly and actively unites psyche and soma, and illuminates cognition;

– the importance of tactile contact, particularly in care, in a highly affective manner, but supported by a precise technique[2];

– the base (the pelvis and its contents) as an essential localization of the Affective, which is obvious if we are dealing with embryology;

– the affective and physiological importance of the three diaphragms (scapular, thoracic and pelvic) always interacting, between which the child develops in utero;

– the constant interactive exchanges between the bony pelvis and the skull, and between the contents of the pelvis and the contents of the skull (brain and membranes) through the enteric nervous system;

– the sensory and affective dimension of intelligence;

1 In descriptions of clinical work, we refer to as the "approacher" the person who performs the act initiating the encounter, and "approached" the person who is thus led to react and take up a position.

2 Haptonomic tactile contact presents precise characteristics: it modifies the tones of the approached in a way that immediately gives them a sense of completeness and security, which establishes trust between the approacher and the approached. The approacher integrates the approached by extending to them (an act that is commonly done with objects). This extension can be global or oriented; it invites the approached to respond by integrating the approacher in return (this act is most often performed non-consciously). According to the scale of affective commitment of the approacher, this contact can be either psychotactile affective-confirming or thymotactile.

– the need to be validated in affective terms in order to be able to unleash the full potential that we carry in ourselves;

– the links between the Affective, muscle tones and the subcortical pathways;

– the decisive importance of the parent–child relationship in prenatal life, at birth and in the first months of life. He therefore proposed an epigenetics clinic, long before this became scientifically self-evident;

– early emphasis placed on verticality in babies.

5.3. Haptonomic pre- and postnatal accompaniment, or preparation of verticality during prenatal life

A brief description of this work makes it possible to understand how verticality is prepared as of prenatal life, beyond the organization of the development of the vertebral column, common to all *sapiens* fetuses, as demonstrated by Anne Dambricourt Malassé (1988, 2022a, 2022b). Pre- and postnatal haptonomic accompaniment consists of affective accompaniment for the parents and their child (or their children in the case of multiple pregnancies). It begins as soon as possible after conception and ends when the child has learned to walk confidently, about two months after their first steps, all-important in their "family story". It is interesting to note how frequently the expression "take one's first steps" is used, always giving an affective and positive tone to the act of undertaking.

Haptonomy is always practiced in individual sessions, i.e. as a couple, and not in a group, because the work is too intimate to be practiced by several people at once. If the father has died, or left early in pregnancy, the mother may choose a third party to accompany her and her child. This person will help prevent the development of a hypertrophied, overly strong and stifling dual relationship, that would be potentially pathogenic, for the mother as for the child. This third party is in a fourth position, since the genitor, or father, cannot be erased from the history of the child and no one can take his place; moreover, it is often a woman who is chosen. It is the same for same-sex female couples: the genitor is present in the child and this cannot be hidden.

It is very important to add that haptonomy must never be imposed, either on the mother or the father, as a new ideology to which every supposedly "good parent" should submit. This would be entirely contrary to our ethic of respecting the freedom and autonomy of each individual!

Haptonomic accompaniment is not opposed to medicalization. We often find ourselves accompanying women whose pregnancy is highly medicalized, or heading for a scheduled cesarean section. In all difficult, even tragic, situations, accompaniment helps the parents and the child to navigate difficulties and challenges together, without them feeling either like they are victims or passive. Provided that the practitioners respect each other, excellent synergy can exist between obstetricians and neonatalogists practicing cutting-edge medicine and also haptotherapists.

5.3.1. *Clinical work*

5.3.1.1. *From the parents' perspective*

The uterus of a woman who is affectively connected to her child undergoes immediate changes. This is why we use the word "womb" in haptonomy, to differentiate it from the uterus as a hollow muscle, empty or refusing the child, and the whole inseparably formed by the contents of the pelvis (including the uterus and perineum), the abdominal muscles and the three diaphragms (scapular, thoracic and pelvic). All of these elements are brought together by affective functioning. They form a whole, changes to which are felt by the child, installed in a house of muscles, fascias and ligaments, of which they perceive even the slightest changes.

When a woman is present for her child, in contact with it, her womb – in being experienced as a welcoming place – is immediately altered: changes in tone, dilation of vessels and hormonal changes that alter the taste and smell of amniotic fluid (an anxious mother or a serene mother do not have the same taste). The mother experiences an immediately perceptible sense of well-being or greater well-being. The child then becomes much more present for the mother, manifesting itself by what appears to be a response: a gyrating movement of its spine, which is always perceptible. In gyrating its future spine, it has its first experiences of what will become verticalization, as a process of humanization. This small movement, echoing the change in the presence of the mother, really gives the feeling that the child is indicating to us that it has perceived what is happening, and is reacting to it. It takes position by mobilizing its vertebral axis. As long as we have not perceived this movement, we know that the mother is not yet really with her child affectively, although she is sometimes convinced of it. This response initiates the session: "It is the mother who opens the encounter, in this extraordinary relationship with her child that is 'inextricable'. It may happen, in very specific situations, that the mother is incapable of doing so; then it is the father and the accompanying person who will help the child to manifest itself, as if it were going to seek its mother by calling on her to make contact. The therapeutic impulses of the child begin before birth".

In a very fine non-verbal dialogue, the child's motor skills become an extremely subtle and precise means of expression for someone who has experience of it. This ability of the child to react according to what its mother is experiencing still amazes us after more than 35 years of practice. Moving does not mean answering, stomping does not mean moving; dancing, swinging and stretching are very different things. It is a very fine but precise language that parents learn very quickly to decode.

The mother can also, without using her hands, invite the child to move upward or downward or to one side or the other. She does so by modifying the tone in her uterus, by means of the subcortical circuits. For this, it suffices that she has the desire to invite her child closer to her heart, or to its father or the accompanier, or that she finds the feeling of laying it in the cradle of her pelvis. The tones change immediately in the region invested affectively by the mother, and it is as if the child is drawn in that direction. She discovers that she can rock her child from the inside, without any external tactile contact.

The vertebral axis is therefore the first place where affective dialogue is expressed and experienced, the first expression of the positive experience of "being together", and its mobilization can be understood as the first non-verbal language. It is in this support of its future verticality that the child in utero takes up a position in the encounter. This place, and this lived experience, are therefore contemporaneous with the discovery of an otherness, the primary condition for the construction of the subject. It is through the use of the spine that the child in utero switches from spontaneous movement to a purposefully organized gesture.

During the accompaniment, fathers[3] learn a number of gestures that allow them to bring comfort to the mother, and thus to the child, while the body changes experienced by women are often arduous day-to-day. For fathers, there are several ways to enter into contact with the child from the outside, playful to a greater or lesser extent, tender to a greater or lesser extent, always with hands that are very light but very present. Together, he parents learn this art of dialogue with the child, well before its birth. Little by little, they build a team of three, bonded by the Affective that will prove essential during major events, such as childbirth and delivery, or tragic events that can occur: announcement of a disability, the need to resort to medical termination of the pregnancy or a pathological event in one of the parents. Fathers are always surprised to feel how sensitive the child is to their manual calls, to their voice and to their mere presence at the mother's side.

5.3.1.2. *From the child's perspective*

The child in utero learns constantly and very quickly, and it decodes almost before its mother the direction of the invitation and actively participates in the movement. It all happens as if it were looking out for any possibility of entering into dialogue. We get the feeling that the child is watching out for what might be a

---

3 With the term "father", we are referring to the mother's other, who is not necessarily the genitor. It is he (or sometimes she) who stays by the mother's side and who commits to welcoming and raising the child with her.

signal, and very much distinguishes between an object, such as a belt, an ultrasound probe and a hand. Even more surprisingly, it will not react in the same way to a hand posed without a process of encounter, and a hand seeking contact, even motionless. This remains inexplicable for now, but future research will no doubt one day provide us with an explanation. The discovery of mirror neurons, which are the neurons of empathy, by Giacomo Rizzolati (2008) at the University of Parma, in 1992, opens up very interesting perspectives. At certain moments, the child in utero moves a huge amount, its arms and legs especially, and the parents look out for these strikes, yet they have no relational significance. These movements are necessary for the formation of the limbs, and can also correspond to the arrival of carbohydrates into the amniotic fluid after the mother has eaten a sweet meal; or to relaxation, occurring after a busy day for the mother, which has caused the child to live in a tense and uncomfortable womb. In any case, these are not favorable moments for an encounter: not all movements by the child signify interaction. These moments of motor agitation can often give rise to a harmful misrepresentation: the mother tells the father "put your hand here, the baby is moving", and suddenly the child calms down completely for several minutes, before then returning to a much more peaceful motricity. This is because the father's hand brings to the mother–child dyad a calm they yearn for, a calm that the mother cannot give to herself. Unfortunately, parents rarely wait for long enough, and the father takes his hand away, disappointed, thinking that his child is not interested in his contact, when it is quite the opposite.

The child very quickly recognizes that it is being presented with a game; it approaches and comes to meet the hand that calls it on one side and then on the other. The mother must always accompany the child from the inside when the father calls it: our goal is not to "make the child move", but to have the triad experience affective exchanges in which each has their place, each confirming the other affectively[4], which enriches each of them in a dynamic that develops in them a feeling of affective security. It is by mobilizing its spine that the child organizes its movements; in no case does it perform somersaults, which could present a risk of getting tangled in its umbilical cord. It is also possible to propose games involving lateral or vertical swinging, or winding around its axis. We then discover that after a few minutes, the child takes over the game and organizes the rhythm, duration, intensity and direction of this swinging game. Of course, the mother accompanies the child non-consciously, because without this it would not be possible, and as soon as she is no longer present for her child, it suspends its movement, but it is the child who leads the game. It is in the eyes of the mother that we see that she felt the child

4 Affective confirmation: a haptonomic act involving tactile contact. It makes the person approached feel a sense of well-being, completeness, trust and affective security, so that they experience a feeling of being well in themselves. Haptonomy posits that in order to unleash the full potential that they carry within them, a human being must be affectively validated as early as possible in their history.

take charge of the movement; there is a small change in her gaze, more interior. Later, we discover that the child, having memorized these games, itself comes to propose it. In a time of very early education, the mother can signal to the child from the inside that she is not available to respond to its proposal. The visionary Erasmus (1466–1536) said that education begins before birth.

The child also learns to recognize the sounds and voices surrounding its mother. It gets used to familiar noises, which no longer startle it; Japanese studies have demonstrated this, with children who have lived their prenatal lives near an airport. The child distinguishes between different voices; some please it and it moves closer. Others scare it and it freezes or moves away. Fathers do not know this well enough: when they speak at length next to the mother, the child always comes and approaches the side from which their voice is coming. Until at the beginning of the 3rd trimester, the child does not hear via tympanic vibration but reacts to its whole self being caused to vibrate by a voice that approaches and speaks for long enough; this is why we consider the endomesoblast as the tissue of high affective memory. The taste of the food or products that its mother ingests, the smells she breathes, the sounds that surround her – these rhythms of life become familiar to it to the point that they form part of the child's world before its birth, since it does not yet feel differentiated from its mother. These sensory experiences will leave deep traces over a very long period of time (Busnel 1993, 2021; Schaal et al. 2004). Here again, the spine is concerned because the vibrations are felt there very intensely. Thus, all congenitally deaf children have perceived, via their spine, the voices of their parents in prenatal life. We can therefore consider that the voices perceived in prenatal life form part of the early process of verticalization, genetically programmed and activated by affective exchanges.

Giving meaning to what they live is an essential quest of human beings, from their conception to their death. And it is through the prism of the Affective that each individual gives meaning and significance to the experiences that life offers them. The child, as soon as it is contacted affectively – which means that we address it as a subject – becomes animated, in the multiplicity of meaning that these words resonate in us. This "animation" is accompanied by movements coming from the spine, which begin in its base. Hence, at birth, babies who have been well accompanied hold their head and spine better: they have more experience of it than babies (fetuses) who have not been moved by a desire for exchange have. This is why active exchanges prepare for easy verticalization, specific to humans, as well as access to conscience and the values of humanity.

Our long practice of pre- and postnatal accompaniment has allowed us to keep track of children who have grown up. The evidence of the richness of what they have acquired in their prenatal life and in the first years of life remains and undoubtedly influences their journey. The way they conduct their lives confirms the importance of affective security, of this early and joyful inscription of the chain of lived experiences, of the words spoken and of the cascades of organization of meaning that follow, and remain inscribed in the flesh. Affective security brings security in the way a person conducts their life.

5.4. The endomesoblast: the embryological foundation of verticality

The tissue unit that resembles the substrate and the vehicle of affective information, both locally inscribed and actively circulating, raises questions. This is when reflection on embryology takes on its full strength. Indeed, there does exist an elective tissue of affective expression: the endomesoblast. The spine and what surrounds it are mainly derived from it.

The embryo at less than 20 days old is a tiny disc, formed of three layers. From the outside to the inside, we find (see Figure 5.1):

– at the top, the ectoblast, which will give rise to the epidermis and the nervous tissues;

– in the middle, the mesoblast, which will give rise to the skeleton, muscles, ligaments, tendons, fascias, all connective tissue, the circulatory system with the heart and lungs, the renal system with the adrenals and the urogenital system;

– the last layer, the endoblast, which will give rise to the digestive glands, and the digestive and respiratory epithelia.

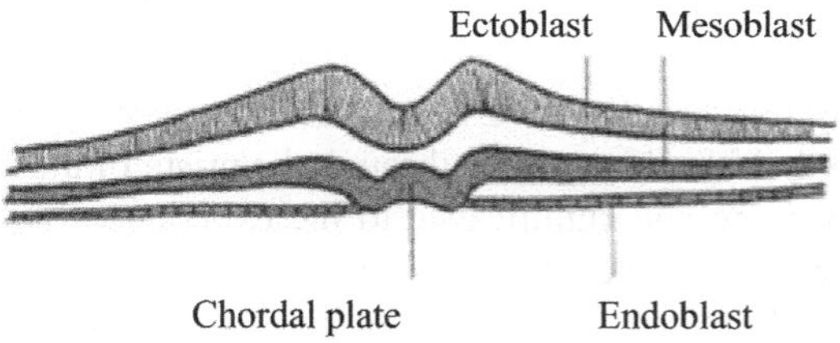

Figure 5.1. *Cross-section of the dorsal chord according to Truchman-Duplessis (1968); courses of the the Frans Veldman International Centre for Research and Development in Haptonomy*

These tissue layers are the origin of essential organs. At 21 days after conception, the primitive heart begins to beat.

The mesoblast is like a sheet composed of bones, muscles, tissues and organs, glands, a source of hormones and blood vessels that carry them. Along with vascular variations come the changes in muscle fibers under the effect of neuromuscular spindles, which Frans Veldman described in detail in his book. This tissue is extremely reactive to emotions and affective feelings. At the slightest feeling of insecurity, fear or refusal, it retracts and stiffens, whereas it expands and softens as soon as affective security, acceptance and openness are there.

This happens as of prenatal life (see Figure 5.2). The fact that it forms the musculovascular network that runs from the feet to the head makes it the emotional tissue par excellence, since, inter alia, the vessels transport hormones. In an instant, moved by an emotion, a feeling, a thought or a word, the muscles and vessels react, at the same time as the breath changes.

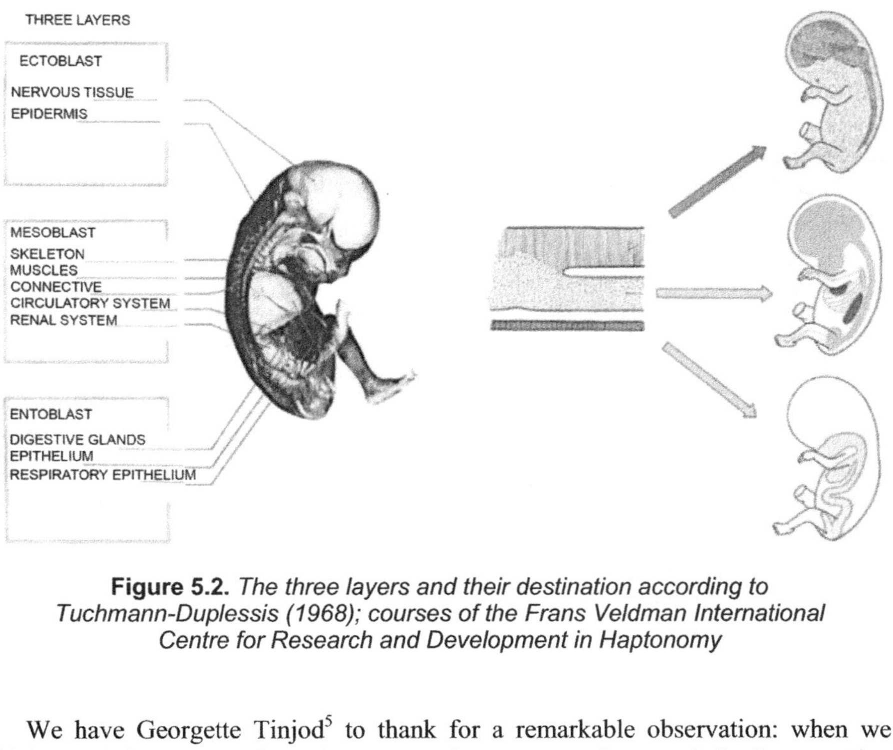

Figure 5.2. *The three layers and their destination according to Tuchmann-Duplessis (1968); courses of the Frans Veldman International Centre for Research and Development in Haptonomy*

We have Georgette Tinjod[5] to thank for a remarkable observation: when we think strongly to ourselves, in our mother tongue, the word "no", a tension immediately occurs in the lower back, at the Michaelis rhombus, which then spreads to the base, then throughout the person. A locking movement takes hold throughout

5 Georgette Tinjod, a midwife, has been a member of the teaching team of the Centre international de recherche et de développement de l'haptonomie (CIRDH-FV: International haptonomy research and development center) since its founding.

the spine, from the bottom upward. This happens as soon as we are upset, angry, frustrated or even simply seized by doubt. But very few people know how to feel this; they content themselves with feeling tense, uncomfortable or having a backache [...]. Conversely, when we think strongly of the word "yes", a relaxation in all these regions occurs and then diffuses throughout the corporeality, and is experienced as a state of well-being, openness to the world and security.

We can thus seek with each and every individual the words that put them in a state of either non-receptiveness or openness, i.e. refusal or acceptance. In critical situations, such as the threat of a premature delivery, this can be the key to a happy outcome.

This observation, which seems quite simple, is in fact very important, because it is a way for us to understand how signifiers and experiences, whether good or bad, are inscribed in the flesh, via the base and the endomesoblast, and leave their mark there. Thus, the subject's intimate and affective history is engraved in the most secret part of our being, such as a tattoo. And it would seem that it is the two axes, anterior and posterior (see section 5.6), that are the most active in this reactive dynamics.

The adrenal glands, with their production of adrenaline and cortisol, as well as the urogenital system, are also of mesoblastic origin, which reflects the importance of experiences in the base. There is also the production of serotonin by the enteric nervous system. Frans Veldman insisted that one of the effects of thymotactile contact[6] was the production of serotonin. These tissues, having a single origin, have a continuity, a particular proximity and a unique reaction temporality. They are all extremely sensitive and reactive to affective contact. When we enter into tactile contact, it is beyond the epidermis, the endomesoblast, that is contacted. The epidermis originates from the ectoblast. It is this *continuum* between the three layers that explains the feeling of completeness brought by a high-quality contact, whether psychotactile or thymotactile.

It is for this reason that haptonomy offers a unique and effective approach in panic situations when a powerful hormonal flood rises from the enteric nervous system, too massive for the cortex or the cortical innervation via control of the thoracic diaphragm, to be able to curb it.

Certainly, our sense organs come from the ectoblast. It constitutes the fabric of relations with the outside world, but without the link with the support tissue derived from the mesoblast, the sense organs could not nourish intelligence as conceived by

6 See note 2.

haptonomy (*hapsis*[7]), nor engage in their exchanges of sensory modalities, which play such a significant role in the construction of meaning and discernment. It is the space where the original synesthesia unfolds, which is so important for the construction of the subject. We refer you here to the theory of Françoise Dolto (1984) on "the unconscious image of the body".

It is the mesoblastic tissue that connects within the body space what comes from one part or the other, from one direction or another. It thus allows two events occurring far from one other to be immediately connected, in the space of the flesh, to make sense, to give a feeling of insecurity or security, or of completeness or incompleteness. It is the endomesoblast that connects, in the continuity of the time of our history and in the fleshly space, the embryo, then the fetus, then the child that we were, and all of the person that we have become.

The endoblast is also of paramount importance insofar as it will produce the epithelia, which are the places of alchemy where we draw our vital elements from what is given to us from the outside, whether oxygen or food. The respiratory epithelium is of extraordinary affective importance: the first breath occurs at the same time as the immense upheaval that is birth. Its importance is existential: without breathing, there is no survival. The digestive epithelium also represents a particular singularity insofar as it is strongly involved in the early parent–child relationship. The digestive tract is the place where parents feel they can legitimately control what goes in and out, with an intrusiveness that is sometimes staggering, linked to many narcissistic feelings. If the child eats well and relieves itself well, it is because we are good parents: requiring it to eat and monitoring its bowel movements are indisputably parental obligations. We have heard it be said for centuries, and it is not entirely false!

One final essential point: the neural tube closes from the head to the sacrum, and the closure of the bottom occurs 28 days after conception, two days later than the top. It is at this moment and in this place that the endoblast and mesoblast become joined to form a *continuum* reacting affectively as a whole. It is on their meeting that unity depends, which makes it possible to understand the transversality of our perceptions, those of the approacher as well as those of the approached.

This explains the feeling of completeness brought by affective contact (always chaste) in the base region. All this, added to the anatomical and physiological reality, allows us to understand the importance of the base, the place where tissues culminate and gather, the place from which we hold our head, and adjust our

7 *Hapsis*: the matrix of the intelligence of life; it is the basis of discernment. It constitutes the filter by which the sensory becomes sensual, charged with meaning. It organizes the interconnection between body and mind at every moment.

posture, on which Frans Veldman has always placed great emphasis. The base is an essential place for affective life, much more than for sexuality, as is still too often overlooked.

5.5. Epigenetics and neural plasticity

We now know the importance of epigenetics. The environment, in the broadest sense of the term, is in an ongoing interactive dialogue with genetic heritage. With any given identical genetic data, certain genes will be expressed, while others will have their expression inhibited or modified depending on what the child experiences in the womb and during early childhood, with significant reshaping around birth and in the early postnatal period. In fact, this dialogue continues, while less intense, from conception to death. We are thus freed from the supposed tyranny of genes, but we understand how great the responsibility is of all those who care for parents and children in what are such sensitive periods. Similarly, the term "neural plasticity", introduced by neurologist Geoffrey Raisman in 1969 (Huang et al. 2015), allows us to understand why babies who are well accompanied before birth have an intense presence and a keen gaze, while at the same time being very peaceful and confident. This cannot be separated from the fact that their posture tone is better: they have a good axial tone, hold their head better and their limbs are less hypertonic. Neural circuits have been created by exchanges, the "communicative apparatus" as a whole has developed, because it has been called upon to function, the spine has been invested, inhabited and used in the movements that the child makes to respond. Proprioceptivity has already been exercised before birth, with the neural stimulation that this implies. Memorization, learning to identify sequences: the bell rings, steps approach, a presence and a voice arrive: the child has already reacted according to what this presence represents, good or bad, for its mother and itself.

5.6. The two axes of humanity

Verticality is consubstantial with human beings, even for those who will never stand on their legs. There exists a lived experience of the vertebral axis whose dynamics extends from the sacrum to the occiput, which any human being supported at their base, even when sitting, or lying down, can experience as reassuring and energizing. To put this axis in motion is to take up our place in the world, to experience the fact of being ourselves, individuated.

When a human, whatever their age, is sat upright on their sacrum (if we may be allowed this barbarism), their eyes are in front of the holes, their brain functions in another manner, they actively take their place in the world, as seen in the photos in Figure 5.3.

Figure 5.3. *One-month-old baby resting on its father's hand (source: Dr. C. Dolto)*

The age-old story of humans finding something specific to their humanity in their verticality is replayed for every child. The discovery of verticality is an aesthetic emotion, organizing and humanizing, strictly speaking. The act of birth implies a very strong push directed from the sacrum toward the occiput. The newborn, born naturally, vaginally, is full of the power of this push that they have just experienced, of this force that has mobilized its spine. To resituate this to the child is to acknowledge this power in it, to allow it to appropriate this power in order to be capable of taking its place in the world: its full place, but nothing but its place. Enriched in its past to slip into the present, to better shape its future. For those who were unable to experience it, because of a cesarean section, the discovery of verticality is even more important. I refer to the posterior axis, that of the spine, as "the axis of autonomy", because it is from its mastery that the ability to have distance from adults originates. This essential axis is in constant dialogue with the anterior axis, the soft axis of the digestive tract and respiratory tree, of opposite polarity, from top to bottom thanks to the peristaltic movement. The digestive tract, throat and respiratory tree tighten as soon as the child, or the adult, feels in objective or subjective insecurity. This is why we call it "the axis of alienation", that in which are inscribed dependency, anxiety over unmet needs and fear of abandonment that accompanies deprivation, missed appointments and misunderstandings (the child cries because it needs to be in contact, or to exchange, but it is given food. Or, the child cries out of hunger, but is spoken to reassuringly without being fed).

These two axes of reverse polarity, animated by a discrete but constant mobility, are in permanent interaction. If we understand them as such, clinical practice is often enlightened. To stand up is the desire that puts these two axes in a permanent, dynamic, and sometimes contradictory relationship.

5.7. Another to stand up straight

For a toddler, feeling upright, standing if it is big, sitting if it is a baby or premature, with its spine resting securely on a hand that is in constant dialogue with its occiput, signifies to it its entry into humanity. This has a sensual significance, even stronger if accompanied by words that give it meaning. It is the living nature of the hand that supports it at its base, which leads it to discover that it has the possibility of holding its head up at the end of its spine. The same child, sitting on a piece of furniture or placed in an inert hand, will let its head drop onto its spine, failing to maintain verticality.

The first postnatal discovery of verticalization is thus given by someone similar, a fellow human, who offers the subject the possibility of discovering what is within itself, which it cannot have without an encounter with another. This point is essential because it roots the first verticality in an exchange of mutual trust. Everything changes; there is a before and an after. The perception that the child has of the world and of itself profoundly changes it. The philosopher Emmanuel de Saint-Aubert (2016) proposes the concept of "lift", which brings together in a single word the entire humanizing ethic of an action that allows the child to draw on its own resources, because it is supported in the multiplicity of meaning of the term.

5.8. Verticalization and postnatal work

Children who are well accompanied affectively require us to rethink pediatric semiology and the stages of development. We propose a specific way of carrying and accompanying them in the first months of their lives. This represents a logical progression, since for them, the ability to move freely in utero is certainly a difficult loss. Moreover, in prenatal life, tactile contact is at the forefront. For months, sounds are perceived through the skin and endomesoblast; the child plays with the walls, rubbing against them; it touches its placenta with its legs and feet, licks it, manipulates its cord, sucks its thumb and masturbates. Therefore, during this period, tact is its primary instrument in discovering its own world. At birth, it discovers gravity, motor incoordination, fear of heights and dependence; and because it is totally vulnerable and unable to meet its own needs, this is undoubtedly the origin of the fear of abandonment, which will remain very significant for the vast majority of humans. There is certainly a great lack of tactile contact that requires it to open up other channels to develop its intelligence in the face of sensory chaos, and a need to adapt to new and likely very difficult living conditions.

It is therefore a question of carrying the child, and at each age giving it the opportunity to feel secure, and to feel all the autonomy it can have at that point in its development, however minimal that may be. It is the feeling of autonomy more than its reality that is essential. This is experienced simultaneously with a feeling in the vertebral axis.

We ask the father, where medically possible, to be the first to separate the mother from the child, by having it live its first experience of verticality without being enveloped. If he cannot do this in the birth room, he should do so as soon as possible once they are alone and the child is available, i.e. neither hungry nor asleep. This initial verticality without being enveloped, which the child experiences as it looks at its mother for the first time, allows it to reform its triad of security; this is a great founding aesthetic emotion. When carried in Buddha pose, with their legs bent, their sacrum resting on the wearer's sternum, facing outward, but well supported in their base, newborns and those born prematurely seek their verticality. In a movement that starts from their base, they lift themselves up and hold themselves upright for six to seven seconds. Their gaze is very present – they observe everything around them – then after six to seven seconds, they lean back again, and their gaze withdraws, before they again straighten up after a resting period of six to seven seconds. This alternation allows parents to feel that their child is seeking its verticality and that it fully feels the moments that are favorable to opening up to the world, with its spine active; and moments of rest, with its spine resting. Together, the parents and child learn this subtle attunement between an appropriate proposal and an empowering reaction.

We teach parents how to always give the child both its feeling of completeness, and the feeling of its vertebral axis from the base upward. Through what the child discovers, through the place it occupies and through the way its gaze can call on others, the verticalized child finds itself a part of its own kind's group. The most difficult aspect for us is to have parents understand that it is from the base that we hold a child's head and that, conversely, by holding its head, we lead it to believe that it cannot do it itself, when it is perfectly capable of doing so, briefly at first, then for longer and longer periods. The moments where the child is picked up or laid down are very important, if parents are taught to do so in such a way that they do not need to hold the head, allowing the child to use the lateral muscle chains in its neck. The child learns very quickly to take charge of its spine, which gives it a sense of security and autonomy. At a few weeks old, supported by the base, reclining on the bed, its brain very active, the child straightens its head and back effortlessly and looks at its parents with the intense presence of a child who feels it is mastering its verticality as a confirmation of its being, as a promise of what it is to become.

It is striking to witness the deep joy, imbued with solemnity, of the first vertical position; it is the only discovery that allows a baby to forget, for a few minutes, that it is hungry. Standing vertically is an exciting act for someone who is discovering it. Faced with the sensory chaos that surrounds and animates the newborn, we believe that verticality is pacifying in that it organizes chaos and centers the child in its action.

We then see them again between the age of three and four months to evolve the carrying style: if a child aged over four months is carried like a three-month-old, it gets bored. Then, we see them on the "anniversary" of their birth[8], nine months for a child born full term. This is a very sensitive time, during which the child and its parents potentially experience anniversary syndrome regarding the birth, delivery or challenges that preceded or followed the birth. We like to call this session the celebration of their "*arriversaire*[9]". Often, the child is already seeking to stand up to walk. The last session takes place when the child is able to walk confidently, two or three months after mastering walking. The "bipedal revolution" brings about considerable changes in the sense of self; it imposes an educational revolution for these very secure children who, if not contained, could become too powerful affectively in relation to their parents.

Access to walking imposes the beginning of a consciousness of the art of living together and tolerance of frustration, without which no social life is possible.

5.9. Conclusion

In the beginnings of life, the desire for verticalization always prevails for a few minutes over the need for food. This should give us cause to reflect on the profound meaning of verticality and allow us to understand how it can be linked to uprightness, i.e. to the values of humanity. The transcultural, and the universal, are manifested through the singular, which is to be constructed each time, at each birth. For all of us, it involves shared experiences, of which verticality is an essential element. This is what leads us to say that there is something sacred in this joy of verticality, something that attains the universal, that touches us in what Jacques Lecoq[10] called "our common poetic foundation" (Lecoq et al. 1998).

In all cultures, since the dawn of time, rituals and practices have been found to signify to the arriving specimen of the species its belonging to a group, an ethnicity or a clan.

8 In fact, it is a time period equal to that from conception to birth.

9 French term combining *anniversaire* (birthday) and *arrivée* (arrival).

10 Jacques Lecoq (1921–1999), man of the theater, great teacher, founder of the École internationale de théâtre Jacques Lecoq (Jacques Lecoq international theatre school).

Nowadays, the human species has become so pretentious that it behaves as if it were enough to be born of a man and a woman to develop like a human. We only need to open our eyes and ears to see that this is not true. The small human mammal needs to be humanized.

In the Middle Ages, the child was enclosed, with its arms along its body, in tight swaddling. This practice was largely intended to give the child a vertical form and to enable it not to remain in the larval, curved form of the newborn, associated with animals and the image of the devil. Verticality was then clearly an indication of the victory of humans over an animal nature potentially linked to the diabolical, with the impulsive thus on the side of evil.

Representations of virgins and the children of Jesus, until the end of the 12th century, stand upright, in their base, hieratic, with their "eyes in front of the holes". It is only beyond the 13th century, when the crèche was promoted and Christianity presented as a family adventure, that children are shown to be languid, and slouching. They have lost their verticality, the virgins have become charming, displaying a fulfilled mother's smile (see Figure 5.4), but the dolorism is there in the Christs whose hips sway; the symbolism of verticality and uprightness has disappeared.

Figure 5.4. *Drawing of a Virgin in Majesty statue, 13th century, Burgundy, France (by Chloé Hakoun)*

Giving access to verticalization is a call made to the child. It is a call to its humanity, which is within it, but which must be called upon for it to awaken. Doing so during prenatal life leaves traces that will foster the transmission of this humanity and the development of the child's personality.

5.10. References

Arendt, H. (1983). *La condition de l'homme moderne*. Calmann-Lévy, Paris.

Busnel, M.C. (1993). *Le langage des bébés*. Éditions Jacques Grancher, Paris.

Busnel, M.C. (2021). *L'aube de l'être, souvenir des bébés*. Langage, Paris.

Dambricourt Malassé, A. (1988). Hominisation et foetalisation. *Comptes Rendus de l'Académie des Sciences*, 307(II), 199–204.

Dambricourt Malassé, A. (2022a). *Embryogeny and Phylogeny of the Human Posture 1: A New Glance at the Future of our Species*. ISTE Ltd, London and Wiley, Hoboken, NJ.

Dambricourt Malassé, A. (2022b). *Embryogeny and Phylogeny of the Human Posture 2: A New Glance at the Future of our Species*. ISTE Ltd, London and Wiley, Hoboken, NJ.

Dolto, F. (1984). *L'image inconsciente du corps*. Le Seuil, Paris.

Huang, H., Raisman, G., Sanberg, P.G., Sharma Hari, S. (2015). *Neurorestoratology: Clinical Progress of Neurorestoratology*. Nova Science Publishers, New York.

Lecoq, J., Carasso, J.B., Lallias, J.C. (1998). *Le corps poétique, un enseignement de la création théâtrale*. Actes Sud, Paris.

Mondzain, M.J. (2023). *Accueillir. Venu(e)s d'un ventre ou d'un pays*. Les Liens qui Libèrent, Paris.

Rizzolati, G. and Cinigaglia, C. (2008). *Les neurones miroirs*. Odile Jacob, Paris.

de Saint-Aubert, E. (2016). Introduction à la notion de portance. *Archives de philosophie*, 2(79), 317–343.

Schaal, B., Lecanuet, J.P., Granier Deferre, C. (2004). Le fœtus et son environnement : premières manifestations de la perception et de la mémoire. In *La grossesse, l'enfant virtuel et la parentalité*, Missonnier, S., Golse, B., Soulé, M. (eds). PUF, Paris.

Tuchmann-Duplessis, H. (1967–1971). *Embryologie*. Masson & Cie, Paris.

Veldman, F. (1998). *Haptonomie science de l'affectivité, redécouvrir l'humain*, 7th edition. PUF, Paris.

6

New Paradigm for Human Parturition

In tribute to Dr. Christine Berge (1953–2020)
Paleoanthropologist at the CNRS
and the Muséum national d'Histoire naturelle

6.1. Introduction

In the late 19th and early 20th centuries, a series of comparative anatomy studies were conducted on the development and morphology of the pelvis of human and non-human primates. Researchers such as Joulin (1864), Topinard (1891) and Le Damany (1912) stand out among the first to have demonstrated that the adaptation of the human species to permanent and exclusive bipedalism induced significant morphological changes to the pelvis. Joulin notes that in mammals, it is the cranial dimensions of the fetus that can pose difficulties. Le Damany sees congenital hip dislocation in women as a defect inherited from evolution and some, such as Keith (1924), cite prolapses and hernias as inheritors of erect posture. These pelvic transformations, such as the short distance between joints to support the weight constraints of the upright position (Reynolds 1931), are necessary for the adoption of a new locomotion. Next, the gradual increase in brain volume in *Homo sapiens* modifies parturition, and a unique obstetrical mechanics, which does not have any equivalent in other primates (Schultz 1949), develops. No consensus exists as yet as to the consequences, often subsequently considered negative, of adaptation to bipedalism on human obstetric function, an issue that will begin to gain structure in the mid-20th century.

In 1951, Krogman described human parturition as one of the "scars of evolution", positing that the two hominization processes – bipedalism and

---

Chapter written by July BOUHALLIER.

encephalization – are opposed to each other. The hypothesis of a mismatched fetal-pelvic confrontation between the fetal skull and the maternal bony pelvis, characteristic of human parturition, is put forward and would explain the difficulty of childbirth in women, compared to that of females of other anthropoid primate species.

When Washburn, in 1960, coined the term "obstetrical dilemma", he referred to the antagonism of morphological needs necessary for each function, which would have created a compromise on the pelvis. This compromise would have been resolved by the prematurity of the human fetus. The obstetrical dilemma is "adopted" by anthropologists and is traditionally defined as the close confrontation between the size of the fetal skull and that of the maternal pelvis, explaining dystocia and the associated high maternal mortality.

Obstetric function in *Homo sapiens* is thus described as complex and also particularly dangerous for our species. In 1988, Trevathan saw the assistance of women during childbirth as a necessity in the evolution of human parturition.

According to the authors, maternal mortality due to osseous dystocia[1] varies between 2% and 8%. Although deaths in human childbirth are more frequent than in present-day, non-human primates, one study nevertheless reports maternal mortality rates as high as 48% in Saimiri monkeys (Rasmussen et al. 1980). In other mammals, mortality rates equivalent to those of humans are reported, and certain domesticated species show much higher rates of osseous dystocia than the human species.

It should be recalled that girls forced into early marriage are particularly at risk of maternal mortality. The issue of adolescent girls married early remains relevant today: according to recent data, in India and sub-Saharan Africa, about 50,000 adolescent girls die each year during pregnancy, childbirth or as a result of childbirth. It is also crucial to recall the definition of maternal mortality (World Health Organization (WHO)), as the majority of deaths are not related to bone structure. It corresponds to maternal death from the first day of conception to 42 days after its termination. In France, the length of time after conception for a death to be defined as maternal is one year. Between 2016 and 2018, the leading cause of maternal death in France was suicide (Saucedo and Deneux-Tharaux, 2024).

Since 1960, the obstetrical dilemma has become the paradigm of human parturition; it has been reinterpreted numerous times. It is not only responsible for

1 Osseous dystocia refers to the pelvis of a pregnant woman presenting an abnormality in its size, shape or inclination that complicates the passage of the fetus during birth, or a fetus whose skull or shoulders are larger in size than the maternal pelvis.

osseous dystocia but also prolapse, uterine fibroids and preeclampsia. It was disseminated for 60 years in medical and academic communities and was taught in midwifery schools. In the 1990s, other hypotheses were put forward to explain the narrow fit between the fetal skull and the maternal pelvis. Recent analyses now apply an intermediate stage of the obstetrical dilemma to non-human primates. For several years, the dilemma has been subject to criticism, and other hypotheses have emerged, notably the coevolution of maternal pelvis and child's skull, and some research demonstrates that nutritional deficiencies and social and cultural factors play a decisive role in the outcome of childbirth and maternal mortality.

6.2. The pelvis and obstetric function

6.2.1. *The human pelvis*

Occasional bipedalism, which is thought to have emerged about 7 million years ago among the ancestors of *Homo sapiens*, led to a major reorganization of the pelvis, with notable consequences on obstetrical mechanics. Moreover, Krogman (1951) suggests that human parturition played a role in the selection of pelvic morphologies.

In humans, the upper part of the pelvis is characterized by an increased width and a reduced length. In non-human primates, the pelvis is typically elongated in the anteroposterior plane, with flat iliac wings. In contrast, in humans, the pelvic cavity takes the form of a torus segment (Malinas and Favier 1979), and three distinct obstetric planes can be defined within it (pelvic inlet, midplane and pelvic outlet). The human pelvis thus adopts a bowl shape, contrasting with that of non-human primates, notably by the presence of a curved, short and sagittally oriented ilium, as well as long and protruding ischial spines, although lost in the wall of the muscles. The sciatic notch is deeper compared to that of non-human primates. Reduced distances between joints also allow for compression of the viscera and weight distribution (Reynolds 1931; Schultz 1961; Berge and Ponge 1983; Berge et al. 1984; Tague and Lovejoy 1986; Berge 1991).

The pelvic cavity, also called the lesser pelvis, plays a decisive role as a "delivery canal". It is important to emphasize that this cavity is neither a bone nor an organ, but a space delineated by the inner walls of the lesser pelvis, furnished with muscles.

The description of the shortening of the human pelvis, compared to that of non-human primates, deserves particular attention in the context of studies on pelvic morphology. According to Abitbol (1988), the reorganization of the pelvis to adapt to bipedalism led to the need for a less spacious pelvic cavity, aimed at preserving the stability of the pelvic floor. However, Le Damany (1912) evokes in his works the hypertrophy of the human pelvic girdle as an adaptive process. Le Damany points out that the pelvic ring, in response to the straightening linked to bipedalism, allowed a fit favorable to the increase in fetal skull size. This modification facilitated the increase of the brain dimensions of the fetus while allowing an efficient passage during childbirth. From this perspective, the human pelvis has evolved to allow the passage of increasingly voluminous fetal heads, a fit that has been made possible by the transformation of the pelvic structure. Furthermore, Papillault (1902) demonstrated that the enlargement of the angle formed by the anterior face of the sacrum with the pelvic inlet plays an essential role in obstetrical mechanics. This angle, which widens with the elevation of the pelvis, is particularly open in women, and this increase helps facilitate the passage of large fetal heads. This anatomical adaptation, far from being an obstacle, appears to be a key factor in the effectiveness of the parturition process in man.

Yet, Schultz (1949) was one of the first to suggest that the constraints of bipedalism tend to reduce the proportions of the cavity, leading to difficult delivery. He was followed by many authors who presented the terms of sexual dimorphism as pelvic consequences of obstetric constraints on the female pelvis (Leutenegger 1974; Ullrich 1975; Tague 1991) and better male locomotion (Bruzek 1985). Since the works of Waterman (1929) and Schultz (1949), the authors widely recognize that the morphology of the pelvic cavity is primarily shaped by locomotor adaptations (Schultz 1949; Leutenegger 1974, 1982; Abitbol 1987). Certain authors highlight the influence of the environment on pelvic morphology (Wheeler 1991; Fischer and Mitteroecker 2015; Wells 2015; Manfredini 2020) or on the dimensions of the pelvis (Ruff 1994; Weaver 2002). Others (Kawada et al. 2020) suggest the existence of a covariation between the maternal pelvis and the fetal skull.

6.2.1.1. *Variability of the pelvis*

In 1933, American researchers Caldwell and Moloy conducted a radiographic study of hundreds of female pelvises as part of an anthropological analysis. Their main objective was to study the anatomical variations of the female pelvis, in order to better understand the optimal shape and dimensions to facilitate childbirth.

Radiographies allowed Caldwell and Moloy to classify pelvic shapes according to a typological system, which influenced medical and surgical approaches at the time. Their study had practical implications for childbirth planning, particularly with regard to the risks associated with a pelvis shape deemed too narrow for natural childbirth.

According to a study based on the analysis of 3,000 pelvic radiographs, the android type, as described by Caldwell and Moloy (1933), represents about 32.5%

of the female population; with the gynecoid type, meanwhile, present in 41.4%, the anthropoid in 23.5% and the platypelloid in 2.6% of cases. Caldwell and Moloy also point out that in a population of 1,000 pregnant women, about 12% present a true android type, while 30% present a gynecoid type. They also note that a large android pelvis can enable childbirth.

After the studies of Caldwell and Moloy, medical technology improved, notably with the evolution of radiography and measurement techniques. Thoms (1941) and Snow (1952) were able to rely on more precise tools to better analyze variations in pelvis shapes.

Snow (1952), unlike Caldwell and Moloy, considers the platypelloid type a variant of the gynecoid type and the anthropoid type a modification of the android type. In his study of 50 male pelvimetries, carried out following the work of Caldwell and Moloy, Snow observed the presence of male pelvises presenting typically female characteristics. However, he did not provide percentages regarding the distribution of pelvic types within the male population. These pelvic types and their varieties in the female and male populations are corroborated by a morphometric analysis of the pelvic cavity (Bouhallier 2000), which distinguishes two platypelloid types and a single gynecoid type within the female types. Moreover, male "hypertyped" pelvic cavities (android) are reported to be less frequent in the male population than female "hypertyped" cavities (gynecoid) in the female population (Bouhallier 2006). According to Snow, four main factors influence pelvis shape: hormones, nutrition, development and accidents possibly occurring in childhood.

There exists a great variability of pelvic shapes within the human species. Abitbol (1996) states that in addition to genetic and hormonal factors, the exposure of girls to intense physical work in late childhood and early adolescence leads to transformations of the pelvis. According to this perspective, the human pelvis evolves from an anthropoid shape at birth to an android or gynecoid shape at puberty, under the combined influence of an upright posture and the need to contain the viscera (Tague and Lovejoy 1986; Abitbol 1988; Berge 1993).

Although classical pelvic morphological typologies are now considered obsolete, the idea of a typically female pelvis, or gynecoid, persists, it still being seen as particularly suited to the permeability of the fetal skull during childbirth. For Henrion et al. (1994), android and anthropoid morphologies are "flawed", meaning that these morphologies are adverse to "normal" childbirth. Yet, they constitute 48% of the female population (Quereux et al. 1995). Aufderheide and Rodriguez-Martin (1998) note that obstruction of the pelvic cavity is due to abnormal bone structures, consequences of infections or rickets. However, these authors explain that a passage is allowed through the shaping of soft tissues by the fetus. The diversity of pathological pelvic shapes and the variability of non-pathological pelvic shapes sometimes make the line between pathology and normality very thin (Bouhallier 2006). Furthermore, if the pelvis shape is pathological, its function may not be affected (Dastugue and Gervais 1992). Indeed, women with pathologies, such as Pott's disease, can physiologically give birth several times (Bouhallier 2006, 2021).

This research highlights significant variability of the pelvic cavity, including the female pelvic cavity, which is confirmed by other studies (Betti 2017) and makes the selection pressure for reproduction less strict. As a result, this variability explains not only the different fetal presentations but also the multiple types of births observed. Medical interventions during childbirth may also have their origin in this mechanical diversity, which can deviate from standardized models (Walrath 2003; Bouhallier 2006).

It is also essential to emphasize that the pelvis, like any other bone of the skeleton, is the result of a life history and constitutes a synthesis of all the factors that have influenced its shape. It should be recalled that the morphology of the pelvis, particularly the pelvic cavity, continues to change throughout the life of the individual, as several studies have shown (Bouhallier 2006; Waltenberger et al. 2022). Indeed, age-related changes have been observed in the shape of the pelvis in adults, and this morphology can vary significantly after menopause.

Finally, the prediction of dystocic childbirth by the pelvimetric study of women is not relevant since there is no pelvic morphology predicting dystocic childbirth outside of pathologies (Bouhallier 2006); however, a recent study demonstrates that the morphology of the pelvis is associated with the mode of birth (Starrach et al. 2022). The comparison of women who died in childbirth or in the days following childbirth does not show any particular common features of the pelvic cavity nor any similarities with the pelvic cavities of women who died in childbirth whose pelvic cavities were pathological (Bouhallier 2006). The bony pelvis is only one element among other factors allowing a successful vaginal delivery, with no morphology being predictive of maternal mortality. The disparity of differences in cesarean-section rates from one country to the next (Boerma et al. 2018) demonstrates that the use of this operation is not linked to a feto-pelvic mismatch specific to the human species. The decline in cesarean sections in recent years without an increase in instrumental extractions supports this. Finally, analyses with a view to predicting feto-pelvic disproportions do not take account of joint movements. While the platypelloid morphology is described as the least favorable for vaginal delivery, its coxo-femoral amplitude is the strategy for avoiding dystocia; it all depends on the position in which the woman gives birth. The studies are conducted on women who have all received epidural analgesia, recognized to increase instrumental extractions (Aveline and Bonnet 2001).

6.2.1.2. *Human obstetrical mechanics*

Obstetrical mechanics, defined as the study of the evolution of the fetus through the maternal pelvis, was described by Farabeuf and Varnier in 1904. The human species is characterized by an obstetrical mechanism different from that of other primates, thought to result from complex interactions between adaptation to bipedalism (corresponding to a small pelvic cavity) and the phenomenon of encephalization (which would correspond to a voluminous fetal skull in proportion to the body size of the fetus) (Joulin 1864; Leutenegger 1974; Lindburg 1982; Rosenberg 1992; Berge 2003). In humans, the skull must rotate and flex forward so that the birth occurs through the occiput, in order to cross the pelvic planes. The human neonate, after a rotation and restitution, emerges with its face turned towards the mother's back (Farabeuf and Varnier 1904; Malinas and Favier 1979; Berge 1988; Trevathan 1996).

This mechanic highlights specific characteristics of *Homo sapiens* following adaptation to bipedalism. Human pelvic morphology requires the fetus to engage obliquely in the pelvic inlet, followed by a flexion and a rotation. Once this phase has been completed, the fetus must then release its head at the front of the ischia, a process called "ante-ischiatic delivery", whereas delivery is retro-ischiatic in monkeys (Berge 1983, 1988). The fetus then proceeds with a restitution of its head. The oval shape of the human pelvis, notably in the antero-posterior plane, is essential for supporting the pelvic floor, and the muscles that make it up influence this obstetrical mechanics.

Human parturition can be broken down into five distinct stages (Farabeuf and Varnier 1904), when the fetal presentation is the presentation of the crown of the skull (crown first).

The first stage corresponds to the engagement of the fetal skull in the pelvic inlet. Through flexion, the head of the fetus engages its occiput into the oblique diameter, which is made possible by maternal resistance. Farabeuf describes this phenomenon as "reduction by flexion".

The second stage of obstetrical mechanics is descent, during which the fetus progresses through the birth canal.

The third stage consists of attacking the pelvic outlet (without taking into account the midplane, contrary to other descriptions that include this intermediate plane). At this stage, the occiput of the fetus performs an antero-posterior rotation, so that it is facing forward.

The fourth stage of the mechanics corresponds to the passage of the pelvic outlet, the moment when the fetal head enters what is called the "soft pelvis". This process continues with the progression of the suboccipitobregmatic circumference, the smallest diameter of the fetal skull. The forehead of the fetus then passes in front of the coccyx by deflection.

Finally, the fifth stage of human obstetrical mechanics is delivery from the vulva, also known as the "delivery phase". This phase is divided into two sub-phases: phase A, where engagement is made by progression and deflection to the suboccipitobregmatic circumference; then phase B, where the forehead and face delivery occurs through deflection of the perineum.

This mechanics can be modified and presents a great variability depending on the fetal presentation, whether it is a face, shoulders, breech or head-last presentation. These variations are not discussed in this chapter.

When the feto-pelvic disproportion is very close, phenomena such as asynclitism (offsetting of the head relative to the axis of the pelvis), nutation and counter-nutation take place.

These joint mobility processes are sometimes observed thanks to the rhombus of Michaelis, an area visible on the pelvis during childbirth, which is a sign of pelvic mechanics and of the presence of joint movements such as nutation (an anterior tilting movement of the sacrum) and counter-nutation (tilting towards the back of the sacrum). This area of the pelvis manifests itself during childbirth due to the pressure exerted by the fetal head as it moves through the pelvis. The rhombus of Michaelis is visible during the passage of the head, and it is often described as a region that shows that the pelvic bones move to allow sufficient opening of the birth canal. This pelvic mechanism was partially described by Ambroise Paré, a 16th-century surgeon, regarding a woman who had given birth 15 days earlier: "The pubic bone was separated in its middle, by half a finger [...]. The tail bone moves back."

Human obstetrical mechanics has been present for 300,000 years (Rosenberg and Trevathan 2007) but a *Homo*-type obstetrical mechanics existed in *Australopithecus* with constraints on the fetal skull as significant as during human childbirth and a forward-curved fetal trajectory as in the human species (Berge et al. 1984; Bouhallier et al. 2004).

6.2.2. *Birth in non-human primates*

In non-human primates, it has been observed that three main pelvic patterns can be distinguished: that of quadruped primates (macaques and baboons), that of climbers and/or brachiators (chimpanzees, gorillas, orangutans and gibbons) and that of bipeds (humans). A similarity is found in the relationship between fetal size and the proportions of the maternal pelvic planes, particularly in baboons, gibbons and humans (Bouhallier and Berge 2006). This translates into the same need to reduce the cranial dimensions of the fetus and to extend the fetal head backward to be born face-first (baboons and gibbons) or to turn and flex the skull forward to be born occiput-first (humans) to cross the pelvic planes. The study shows that the gibbon fetus encounters difficulties passing through the pelvic cavity, the size of the head being very close to the dimensions of the pelvic planes, because their fetus is proportionally large. This raises the question of whether the longer gestation time in these species contributes to a larger fetal volume. The measurements suggest that shoulder width could also constitute an obstacle to delivery in the *Papio* (baboon) and *Hylobates* (gibbon). Abitbol (1987) notes that cercopithecid neonates present a relatively voluminous skull compared to the narrow pelvis of females. Another factor of difficulty in childbirth, mentioned by Rosenberg and Trevathan (2000), is the great width of the shoulders of the human fetus, compared to other primates. Schultz (1961) describes the shoulders of the *Hylobates* fetus as being placed particularly highly, which could hinder the passage of the fetus through the pelvis. According to a new study (Webb et al. 2024), chimpanzees must also deal with a restricted pelvic cavity during childbirth.

These results (Bouhallier and Berge 2006) suggest that certain primates, such as humans, must adapt the fetal head by orienting its dimensions to those of the maternal pelvis to facilitate delivery.

The course of delivery is different for lesser apes, which are usually born with their face turned towards the front of the maternal body (Bowden et al. 1967; Rosenberg and Trevathan 2000) and which sometimes have a difficult birth (Cho et al. 1985).

In conclusion, human obstetrical mechanics, specific to our species, differ from that of non-human primates, in which obstetrical mechanisms also exist, although these have not been subjected to hominization processes. The study shows that encephalization played only a negligible role in obstetric function (Bouhallier and Berge 2006). The validity of the obstetrical dilemma is thus called into question.

6.3. Human maternal mortality

Written sources from great historical civilizations such as Babylon and Ancient Egypt testify to dangerous births (Farnet-Lepiller 1993; Bonnet-Cadilhac 2005). However, other researchers, such as Schofield (1986) and Mafart (1994), question the significance of obstetric mortality in the Middle Ages and in the 19th century. Archaeological research provides evidence suggesting that maternal mortality was not as high in ancient societies as it is today (Douglas 1983; Mafart 1994).

When maternal mortality is observed in ancient populations, it is often attributed to pelvic narrowing resulting from nutritional stress (Sibley et al. 1992) or early births (Tague 1994). For archaeologists (Mafart 1994), female mortalities for bony reasons due to childbirth are not common. Schofield (1986) maintains that the risk of mortality from childbirth was no greater than that of infections. Indeed, it is difficult to establish the exact causes of death of women buried pregnant because fetal remains in the pelvic region do not indicate childbirth complications. Putrefaction often causes post-mortem expulsion of the fetus, suggesting dystocia and that the mother and child were buried in this way (mother and children of the Neolithic in Siberia – see description by Lieverse et al. (2015)), and yet these deaths are often interpreted as complications of childbirth (Alduc and Blondeaux 2002). These first-millennium burials in France and the woman from the Whitehawk site in England confirm this interpretation (Witcher 2019). Other causes of maternal mortality do not leave any trace. Bony remains in the pelvis of a buried woman are absolutely not an indicator of maternal mortality.

The high maternal and infant mortality observed in the 18th and 19th centuries (Tarnier 1858) reinforces the image of a difficult and perilous parturition in humans (Beauvalet-Boutouyrie 1995; Gelis 1995).

The study of mortality records (Schofield 1986) gives the following percentages: for the period 1800–1849, 9.2 women died per 1,000 births in London; outside London, there was a maternal mortality rate of 5.5 per 1,000 births. These differences in the percentage of maternal mortality between childbirth in cities and childbirth outside of cities correspond to the studies of Tarnier (1858) on confinement and lack of hygiene in large cities. These results are entirely in line with the study by Beauvalet-Boutouyrie (1999) concerning deliveries at the public hospitals of Paris under the *Assistance publique* social-protection system (Bouhallier 2006).

"The black death of childbed", the name given to the puerperal fever epidemics of the 19th century, was not linked to a bone problem. However, this fever, caused by streptococcus transmitted by doctors not washing their hands, is responsible for the disaster of women fleeing, refusing to give birth in hospitals. Unfortunately, the poorest women could not afford the services of a midwife and accessed care by entering an *Assistance Publique* hospital, as was the case in Paris.

Interventions by obstetricians became highly invasive, and, from the 18th century onwards, certain maternity facilities were also centers for teaching obstetric science. Births were training situations, involving the application of forceps, English-style childbirth (side-lying) instead of the chair, the introducing of hands into the uterus repeated by students, or the speeding-up or slowing-down of labor depending on students' schedules (Schlumbohm 2002). These practices for training purposes increased the rate of forceps use to 40%. During the same period, Johan Lukas Boër (1751–1835), professor of obstetrics and director of the Vienna maternity hospital, used forceps at a rate of 0.4% (Schlumbohm 2002). These practices and rates would influence the social and cultural vision of childbirth. The iatrogenic effects of contemporary management of delivery in a maternity facility are well documented.

The results obtained after the adoption of hygiene rules in the delivery ward of La Pitié hospital in Paris, from March 5, 1886, to January 1, 1893, were a maternal mortality rate of 1.82%. Dr. Maygrier (1893), the professor in charge of the ward, having imposed these aseptic conditions for seven years, stated, "I need not say that this figure includes all causes of death, whatever they may be, that many of these causes are unrelated to the puerperium, and that they most often involved women who had already been sick". Maygrier prohibited students from engaging in dissection or pathological anatomy, one of the conditions authorizing them to examine pregnant women. Students and doctors transmitted streptococcus by not washing their hands before examining the women after dissecting the bodies. In the bone collection of Henri-Victor Varnier's (1859–1902) obstetrical museum, which preserved pelvises of women who died in childbirth, 167 pelvises showed pathologies related to rickets or pathologies of the hip and spine, 25 pelvises showed no pathology; 49% of the pelvises were generally narrowed, 33% were transversely narrowed (android and anthropoid in shape) and 18% were flattened from front to back (Bouhallier 2000). Certain pelvises bore the name of the maternity facility where the woman died, and by accessing *Assistance Publique* death registers, it was possible, using the dates of death, to collect data such as their age, marital status, occupation and, above all, cause of death. These women belonged to the disadvantaged classes and carried out difficult jobs in unsanitary conditions; the majority were single, and many were born to single mothers. From 1874 to 1876, 73% died from infections, of which 59% were puerperal fever. From 1889 to 1903, the rate of deaths due to infection dropped to 36% due to asepsis rules. The death registers only mention three maternal deaths due to fetal-pelvic incompatibility (Bouhallier 2021).

According to WHO data from February 22, 2023, about 287,000 women died worldwide during or after pregnancy or childbirth in 2020. In 2023, about 70,000 annual deaths are attributed to severe bleeding during childbirth. Of all global maternal deaths, 289,000 deaths occur each year, including 179,000 in sub-Saharan Africa. Among these deaths, 90,000 are concentrated in only two countries: India and Niger (INED 2013). These deaths represent an average of nearly 800 deaths per day, all caused by preventable factors related to pregnancy and childbirth. Among these deaths, it is estimated that about 87% (or 253,000) occurred in sub-Saharan Africa and South Asia, with a marked geographical distribution: about 70%

(or 202,000) of maternal deaths occurred in sub-Saharan Africa, while 16% (about 47,000) were recorded in southern Asia.

In 2020, one maternal death occurred almost every 2 minutes worldwide. However, between 2000 and 2020, the maternal mortality rate, defined as the number of maternal deaths per 100,000 live births, declined by about 34% worldwide. Nearly 95% of maternal deaths were observed to be in low- and middle-income countries in 2020.

Maternal deaths are attributed to a total of 14 groups of causes, which are grouped under seven main categories: abortion, embolism, hemorrhage, hypertension, septicemia, other direct causes and indirect causes (Chong et al. 2024). Osseous dystocia is responsible for 2% of maternal deaths, and it is estimated that 15 million adolescent girls are pregnant each year. Girls under the age of 15, compared to women aged 20–25, have a 5-fold risk of dying during childbirth (De Bernis 2005).

According to available data, about 94% of girls in Bangladesh are married before the age of 18, 27% of them marry between the ages of 12 and 14, and 2% before the age of 11 (UNICEF). As regards maternal deaths in 2016, 38% occur on the day of childbirth, 18% between 7 and 42 days after childbirth and 11% in the first trimester of pregnancy. The two main causes were hemorrhage (30%) and eclampsia (23%); 21% of maternal deaths were associated with other indirect causes, 5% with abortion-related causes and 3% with obstructed or prolonged labor, without distinction (Hossain et al. 2023). This means that the proportion of these deaths involving osseous dystocia is unknown.

Global data relating to causes of maternal death in 2014 reveal the following distribution of the main causes of death: hemorrhage accounted for 27.1% of maternal deaths; hypertensive disorders 14.0% and septicemia 10.7%. The remaining maternal deaths were attributable to abortion (7.9% of deaths), embolism (3.2% of deaths) and all other direct causes of death (9.6% of deaths) (Say et al. 2014).

In Tanzania, the proportion of maternal deaths among adolescent girls aged 15 to 19 represents 13.60%, a figure twice as high as that observed among women aged 40 to 49, where the proportion is 4.80% (95 deaths out of 1,987). It should be noted that no data are available regarding maternal deaths among girls under 15, nor on parity among girls, which prevents a correlation to be established between birth spacing and the risks of maternal death among young mothers (Bwana et al. 2019).

Dystocia, or abnormally prolonged labor, is a complication that accounts for nearly half of cesarean sections in low-risk nulliparous women, due to slow dilation of the cervix. According to a study by Kissler and Hurt (2023), several factors can be responsible, such as uterine contractility problems, maternal obesity, cephalopelvic disproportion or maternal stress. However, the underlying causes of this dystocia are often little studied, and there is a lack of research on the precise justifications for cesarean sections as well as systematic post-cesarean section pelvimetries.

The most significant cause of maternal mortality in *Homo sapiens* is not osseous dystocia, and the latter is not the main reason for cesarean sections being performed. When osseous dystocia occurs, it may involve poor fetal presentation. The etiology of dystocia can rarely be established, and feto-pelvic disproportion cannot be recognized with certainty by measurements of either the fetus or the pelvis. Indeed, data collection is difficult, especially in countries where there is no perinatal routine (Bouvier-Colle et al. 2001). It should be noted that while the majority of maternal deaths have direct obstetric causes in the countries serving as a reference base, these causes are not bone-related. For the quantification of obstructed labor, the data are derived from labor duration and cesarean sections. The observed rate of cesarean sections among women in West Africa is 1.3%. The results, combined with those from published works, suggest a range of 3.6 to 6.5% (median of 5.4%) (Dumont et al. 2001).

In conclusion, it is essential to take into account two major issues in the analysis of maternal mortality: firstly, the precise definition of maternal mortality, and secondly, the impact of social and cultural factors on this issue. According to the WHO definition, "maternal mortality refers to the death of a woman from any cause related to or aggravated by pregnancy or its management (excluding accidental or incidental causes) during pregnancy or within 42 days of termination of pregnancy".

Cultural factors of gender inequality have a considerable impact on women's reproductive health (early marriage, undernutrition and genital mutilation). Maternal mortality rates in certain countries may be directly linked to girls being married before their bone maturation is complete.

6.4. Maternal mortality due to dystocia

Dystocia affects different animal species with varying incidences: 4% in thoroughbreds and 10% in heavy horse breeds, with a low maternal mortality of 0.5% (McCue and Ferris 2012). In sheep, dystocia affects about 5% of births.

Maternal mortality in baboons and gibbons, as well as delivery difficulties in some species of American monkeys, such as the Saimiri, are documented (Bowden 1967; Leutenegger 1982; Ridley 1995; Farber et al. 1997), with a maternal mortality reaching up to 48% among Saimiris under certain conditions (1980). A recent study (Pinka et al. 2024) demonstrates that, despite a narrow feto-pelvic space, Japanese macaques seem to present a lower risk of maternal mortality than humans.

An epidemiological study carried out in Great Britain in 2017 on dystocia in dogs revealed an overall prevalence of 3.7% among 18,758 whelpings. The most affected breeds are the French Bulldog (20.6%), the Boston Terrier (18.8%) and the Pug (14.5%), while the prevalence in crossbreeds is 1.3%. Among miniature dogs, the prevalence of dystocia reaches 25.8%.

Captive reptiles also present a high prevalence of obstructive dystocia, yet they are not viviparous (Divers and Scott 2018).

6.5. The dilemma: paradigm of human parturition

The concept of the "obstetrical dilemma" is central to research on the evolution of parturition in *Homo sapiens*. Introduced by Sherwood Washburn in 1960 in his article "Tools and Human Evolution", published in *Scientific American*, this term describes an evolutionary compromise between the needs of bipedalism and the increasing encephalization of the human species. Washburn suggests that this obstetrical dilemma results from the reduction of the pelvis to allow bipedalism, while selective pressures related to the use of tools favored a larger skull size. The solution to this dilemma would have been premature birth of the fetus, at an earlier stage of development, allowing the fetus to pass through the pelvic canal. In addition, this early birth would have been facilitated by the bipedal position of the mother, who, freed from locomotor constraints, could take care of the still-immature child. Thus, this dilemma would have found its resolution in a shorter gestation and the prematurity of the *Homo sapiens* fetus. However, the obstetrical dilemma is frequently associated with a premise, that of the complexity, difficulty and dangerousness of human obstetric function, such that the dilemma would be a constant compromise on the pelvis of current women.

6.5.1. *The obstetrical dilemma according to Washburn*

Washburn writes:

> But the brain of the fetus must be small enough so that birth may take place. In man, adaptation to bipedal locomotion decreased the size of the bony birth canal at the same time that the exigencies of tool use selected for larger brains. This obstetrical dilemma was solved by delivery of the fetus at a much earlier stage of development. But this

> was possible only because the mother, already bipedal and with hands free of locomotor necessities, could hold the helpless, immature infant (Washburn 1960, p. 74).

For Washburn, the dilemma had been resolved.

6.5.2. *Reinterpretations and criticisms*

However, over the decades, the obstetrical dilemma, resolved for Washburn, has been interpreted as a compromise between the competing selective forces of locomotion and parturition, this dynamic being considered an omnipresent and perpetual factor in the evolution of *Homo sapiens*.

Since 1960, numerous scientific publications have interpreted this dilemma as a compromise inherent to the human species, affecting the shape of the female pelvis, thus explaining the recurrent difficulties of childbirth in *Homo sapiens*, difficulties already present 3 million years ago in *Australopithecus afarensis* (Berge 1993; Wells et al. 2012). Maternal mortality, which is higher in *Homo sapiens* than in other species, is thought to be linked to osseous dystocia, defined as an obstruction to labor due to the mismatch between the size of the maternal pelvis and of the fetal skull (Mitteroecker et al. 2016).

This model of the obstetrical dilemma is widely transmitted in anthropological and medical circles, where it serves to explain not only dystocia but also phenomena such as pelvic-floor lesions, preeclampsia, uterine fibroids and prolapse (Rosenberg and Trevathan 2007; Mitteroecker et al. 2016; González and Chamero 2017; Ruff 2017; Hausler 2021; Vilos et al. 2024; Webb et al. 2024).

The obstetrical dilemma has constituted a major framework for research on human parturition for more than 60 years. However, in recent years, thanks to new methods of pelvis analysis, the concept of the dilemma as formulated by Washburn has been increasingly questioned (Bouhallier 2006; Ponce de León et al. 2008; Wells et al. 2012; Warrener et al. 2015; Dunworth 2021; Wall-Scheffler and Kurki 2022; Kurki 2023). Each author offers their own specific conclusions, but a great number of them, in abandoning the concept of obstetrical dilemma, do not question the causes of maternal mortality or revisit the assumption that human childbirth is inherently difficult and perilous. However, they recognize the impact of social and cultural factors on motherhood. Part of Washburn's hypothesis has been questioned by Falk (1990), who suggests that if the brain started to increase two million years ago, it was not through retroaction between bipedalism and brain size but due to the "radiator" hypothesis. Furthermore, the thermoregulation hypothesis has been proposed to explain certain pelvic dimensions. Wheeler (1991) proposes the thermoregulation hypothesis. Migration and genetic drift are proposed to explain human pelvic diversity (Betti 2014). Wells (2015) associates an ecological link with the obstetrical dilemma and maintains that the potential for maternal size to change over the course of generations prevents the birth weight of her fetus from being subject to a strong genetic influence. Others (Fischer and Mitteroecker 2015) suggest that covariation in the morphology of the fetal skull and the maternal pelvis is the process for improving the obstetrical dilemma.

Wheeler (1991) suggested that adaptations of the human pelvis, notably its reduction, are mainly motivated by the need to regulate body temperature in hot environments, and that the advantage of bipedalism was adaptive protection against heat stress. The adoption of bipedalism led to additional thermal constraints, requiring our ancestors to develop body structures adapted to an environment more exposed to the sun. This adaptation would have favored a narrower pelvis. In the open savanna, quadruped animals expose a larger part of their body surface to the sun, whereas a biped in a standing position would absorb 60% less heat. According to the thermoregulation hypothesis, the efficiency of heat dissipation would have imposed a reduction in the bi-iliac width of the pelvis, and it was therefore thermal selection that restricted the size of the pelvic cavity during the evolution of hominins (Wheeler 1991). The principles of thermoregulation posit a human morphological adaptation to the climate (Ruff 1994).

6.6. Evolution and the midwife

The human fetus must enter the pelvic cavity by flexing its skull to present its smallest diameter to the pelvic inlet. In order to fit to the different diameters of the pelvic cavity, it must perform a rotation and its head is delivered, its face turned towards the back of its mother, preventing her from freeing her child without outside assistance. Trevathan (1988) and Rosenberg (1992, 2005) explain that the obstetrical dilemma is responsible for the invention of the midwife. It is believed that this adaptation to bipedalism and encephalization necessitated the emergence of obstetric assistance. Rosenberg and Trevathan (2000) then posit that the shoulders of the fetus can also cause dystocia. According to Farabeuf and Varnier (1904), the shoulders pass relatively easily, one at a time, through the walls of the true pelvis, as confirmed by Kawada et al. (2022) in a study demonstrating that the shoulders are perfectly adapted for passage through the maternal pelvis.

However, Rosenberg and Trevathan (2018) support the need to limit excessive cesarean sections, highlighting the risks associated with these interventions, both for the mother and the child, and stress the importance of cooperation in the parturition process.

6.7. Recent hypotheses and new perspectives

Other hypotheses integrating the obstetrical dilemma remain current. This is the case with the hypothesis that cesarean section is necessary to relieve the biological cost of human birth (Weiner et al. 2008) or the postulation that increased cesarean section rates in cases of dystocia could lead to an evolutionary "selective pressure", increasing rates of feto-pelvic disproportion by approximately 10 to 20%. This evolution would predict a progressive reduction in the size of the human pelvis under the selective pressure exerted by cesarean sections (Mitteroecker et al. 2016). However, the study of these rates of dystocia is often based on observations made in particular sociocultural contexts, where women's living conditions, such as early marriage and genital mutilation, contribute greatly to obstetric complications. Moreover, estimates of dystocia vary considerably depending on the diagnostic criteria used, making it difficult to establish universal rates (Mitteroecker et al. 2016).

The cliff-edge hypothesis (Mitteroecker et al. 2016; Pavličev et al. 2020) related to the "obstetrical dilemma" suggests that the shape of the human pelvis, having reached the point where it is optimized for bipedalism, is so constrained in terms of its capacity to let through ever-larger fetal heads that any increase in cranial volume could make delivery impossible. The cliff symbolizes this narrow boundary between a birth that is possible and a very difficult or fatal one. As previously discussed, environmental factors influence the morphology of the pelvis (Wells 2017; Wells et al. 2018).

A recent study conducted by Webb et al. (2024) re-examines the obstetrical dilemma through an analysis of feto-pelvic fit in chimpanzees. The simulations conducted reveal restricted pelvic dimensions, particularly at the midplane and pelvic outlet. The authors hypothesize that difficulties related to childbirth gradually developed from an already-narrowed birth canal in anthropoid primates. They posit that obstetric selective pressures have intensified in parallel with the increase in the rigidity of the pubic symphysis, the latter to adapt to the increase in body size. Another hypothesis for an obstetrical dilemma that developed gradually can be associated with research on macaques (Kawada et al. 2020). The alternative proposal is that craniopelvic covariance evolved in the early catarrhines prior to the split of *Hominoidea* and *Cercopithecoidea*, or even prior to the divergence of stem catarrhines, given a high degree of cephalopelvic proportions.

Sexual dimorphism is the subject of a hypothesis in which we find the idea of osseous dystocia being constitutive of the human species. It has been posited that the difference in size between the sexes could be the consequence of a gender hierarchy in which the allocation of nutritional intakes is unequal. Women who are small in size would have a narrower obstetric canal, making the passage of the fetal head at the time of delivery difficult. This gender inequality is thought to have led the species to dystocia and maternal deaths (Touraille and Gouyon 2008). Under this hypothesis, the selection of smaller, slender women in the human species may have resulted in the selection of women with a narrower pelvis. This phenomenon is thought to result from an interaction between evolutionary forces and cultural practices. However, data obtained from pelvic measurements of subjects during autopsies at the Institut Médico-Legal forensic institute (Bouhallier 2015) enable the documentation of specific examples of pelvic-plane dimensions and conclude that no significant correlation was found between individuals' stature and the dimensions of their pelvis nor between their hip width and the size of their pelvic cavity. Kurki (2007) also demonstrates that pelvic dimensions in women of small stature are preserved for obstetric function, which is confirmed by Wells (2015), with the growth of the fetus responding to the maternal somatic and metabolic phenotype. Another hypothesis (Fischer and Mitteroecker 2015) has demonstrated that the dimensions of the pelvis are closely correlated with the dimensions of the fetal skull. Thus, women whose skull is larger, and who are therefore likely to give birth to neonates with a large skull, have a pelvic cavity capable of accommodating these large-headed newborns. Wells (2015) specifies that a woman of small stature with a small pelvis will have rounder pelvic planes, allowing the passage of her fetus.

6.8. Prolapse and large babies

6.8.1. *Stressed motherhood*

Studies have examined the effect of the vertical and lateral positions of the parturient during the second phase of labor on the reduction of perineal tears. These studies on pelvic-floor lesions (Pinka et al. 2024), frequently observed in *Homo sapiens*, show that these lesions are significantly reduced by 80% when women give birth in a squatting position.

The primary purpose of this study was to examine whether vertical (such as standing, squatting or on-all-fours) or lateral birthing postures reduced the rate of perineal tearing compared to the traditional supine position. The results showed that vertical or on-all-fours positions were associated with a lower risk of significant perineal tears.

In contrast, the gynecological position, with feet in stirrups, is not conducive to preserving the perineum. It is suggested that allowing women to close their thighs and allow the child to expel from the back of the pelvis may be more consistent with natural obstetrical mechanics.

It is also interesting to note that certain primates, such as Saimiri monkeys, which are neither bipedal nor highly cerebralized, also exhibit prolapse, calling into question the specifically human obstetric problems believed to be due to the obstetrical dilemma. Indeed, analysis of genital prolapse in a large population of breeding squirrel monkeys demonstrated an association between loss of pelvic support and age and parity. In total, 50% of these monkeys suffer from prolapse (Coates et al. 1996). Pregnancy and childbirth cause lesions on the pelvic-floor muscles and weaken pelvic-organ support in Saimiri females, who also suffer from permanent bladder prolapse after giving birth (Bracken et al. 2011).

Another study (Pereira da Silva et al. 2021) devoted to squirrel monkeys (genus *Saimiri*) demonstrates the impacts of size and poor presentation of the fetus. Thus, squirrel monkeys suffer from perineal trauma, nerve damage and prolapse after parturition of a large fetus.

6.8.2. *Stressed motherhood in humans*

Whether stress is nutritional, social or due to insecurity, its effects are deleterious in all aspects of motherhood: gestation, parturition and childcare.

The association between poor nutrition and an increased risk of maternal mortality has been demonstrated (Manfredini 2020).

Results (Wells et al. 2018) indeed point out that malnutrition causing small adult height and excess weight is likely to have an impact on motherhood in low- and middle-income countries. Nevertheless, efforts to delay the age of marriage, and thus the age of first birth, could at least delay childbirth until pelvic growth is almost complete. Berge (1993) demonstrates that body weight is a major factor in pelvic morphological diversity.

Iron and vitamin-A deficiencies can impair growth rate. Among the consequences of diminished iron and vitamin-A reserves in adolescent girls is an increased risk of menorrhagia, a factor that exacerbates anemia issues in girls who are malnourished during their pre-reproductive years. Moreover, the physiological costs related to growth increase obstetric risks through feto-pelvic mismatch in very young girls who become pregnant (Brabin and Brabin 1992).

Nutritional deficiencies in the father have been shown to cause an “avalanche”: once triggered, it continues to surge even though the initial conditions change (Diniz et al. 2024), propagating deficiencies on a transgenerational level.

Furthermore, the study of the links between large fetuses and dystocia (Alexander et al. 2003), particularly in relation to maternal diabetes, is a rapidly expanding field. It seems that a large fetus is frequently associated with diabetes mellitus in the mother, a factor that could have significant implications for the evolution of the human species in terms of obstetric adaptation and the management of pregnancy-related complications. However, rapid changes in nutritional intake can cause fetal disproportion (Wells 2015).

Finally, scientific progress and the use of antibiotics have played an important role in the decline in maternal mortality (antisepsis and diffusion of sulfonamides) and mark a veritable discontinuity in the trend of maternal mortality, a drastic reduction in maternal deaths due to septicemia and infections (Manfredini 2020). The decline in maternal mortality is often associated with medical care, but it is now recognized that increases in living standards, hygiene and nutrition have played a major role.

6.8.3. *And in non-human primates?*

Maestripieri and Mateo (2009) demonstrate that environmental and social factors, including stress, can affect reproduction and motherhood in non-human primates. The reproductive and maternal capacities of colony-born *Saimiri sciureus* females were impacted by the stress induced by human interactions, notably the withdrawal of the neonate (Taub et al. 1978).

6.9. Conclusion

The obstetrical dilemma did not take place.

Recent scientific discoveries will not allow the old conceptions concerning the obstetrical dilemma to be challenged overnight. These conceptions stem from the vision of Aristotle, Galen and Saint-Thomas Aquinas, among others, for whom women were defective beings, accidental, imperfect males and therefore more likely to fall ill. Within the context of research on the human pelvis, the improvement or evolution of this obstetrical dilemma is discussed, without fundamentally calling it into question. Several elements, resulting from the research previously presented, allow us to reconsider the validity of this concept to describe human parturition.

First, certain research questioning the obstetrical dilemma has highlighted that encephalization has only a negligible role in the evolution of obstetric function, and that birthing mechanisms exist in primates that are not highly cerebralized and not bipedal (Bouhallier 2006; Bouhallier and Berge 2006; Webb 2024). These observations run counter to the idea that the main obstacle to parturition is specifically related to human encephalization.

In addition, covariation between fetal-skull and maternal-pelvis morphology was observed in rhesus macaques, a non-human model that presents characteristics distinct from those of modern humans. These primates, which present neither bipedal locomotion nor a significant degree of encephalization, are sometimes subject to a feto-pelvic mismatch. This suggests once again that specialized pelvic morphology due to straightening and bipedalism, and a high degree of encephalization, are not necessary conditions for the onset of this conflict. This phenomenon may be more widespread than initially thought, the authors conclude (Kawada et al. 2020). Webb (2024) points out that chimpanzees, considered to have a wide pelvis and a small-sized fetus and assumed not to encounter major obstetric difficulties, may also present "tight" feto-pelvic confrontations.

These conclusions show that it is possible that non-bipedal and less-cerebralized species face similar challenges in terms of parturition. The obstetrical dilemma would therefore no longer be a specifically human compromise, whereas exclusive bipedalism and encephalization are.

Thus, the study of these non-human primates calls into question the idea that the obstetrical dilemma is a phenomenon exclusively present in *Homo sapiens*. If this concept can be applied to certain species that have not undergone hominization processes, it now appears that it may no longer be relevant to use it in this context. It seems necessary to reconsider the concept of human parturition, and to place it in a broader context, recognizing that parturition can be risky individually, even in non-human species.

Finally, it is crucial to reassess the percentages of maternal deaths attributed to osseous dystocia. These statistics should not include maternal deaths related to factors such as incomplete bone maturation in young girls, malnutrition in women since childhood or dietary changes related to migration. A clear distinction must be made between cultural and environmental factors and evolutionary processes relative to human parturition. Likewise, it is also important to do so for the populations of the past. It also appears that female pelvic variability is very significant (Bouhallier 2000, 2006; Betti 2017), suggesting evolutionary responses related to other constraints than obstetric ones.

Wall-Sheffler and Kurki (2022) provide additional arguments that different pelvic morphologies are successful in balancing population history, selections and genetic drift.

The concept of a "multifunctional pelvis" (Warrener 2015) seems to reinforce the idea that the pelvis is a "Swiss army knife" of evolution, without doing justice to the remarkable physical and physiological adaptation of the human childbirth process.

The paradigm of the obstetrical dilemma, proposing an explanation for the evolutionary obstetric process in *Homo sapiens*, needs to be reviewed. A few years ago, we were already discussing the perfect match between the fetal skull and the maternal pelvis. It now seems necessary to reorient our thinking and to no longer speak of the "obstetrical dilemma" but of "human obstetrical congruence".

Indeed, it is difficult to deny that human parturition represents a veritable evolutionary success, both from the point of view of biological adaptation and the physiological complexity implemented to enable birth.

6.10. References

Abitbol, M.M. (1987). Evolution of the lumbosacral angle. *Am. J. Phys. Anthropol.*, 72(3), 361–372.

Abitbol, M.M. (1988). Evolution of the ischial spine and of the pelvic floor in the Hominoidea. *Am. J. Phys. Anthropol.*, 75, 53–67.

Abitbol, M.M. (1996). The shapes of the female pelvis: Contributing factors. *J. Reprod. Med.*, 41(4), 242–250.

Alduc-LeBagousse, A. and Blondeaux, J. (2002). Mortalité maternelle et périnatalité au premier millénaire à Lisieux (Calvados, France). *Bull. Mém. Soc. Anthropol.*, 14(3/4), 295–309.

Alexander, G.R., Salihu, H.M., Pass, M.A., Boulet, S.L. (2003). Macrosomic births in the united states: Determinants, outcomes, and proposed grades of risk. *Am. J. Obstet. Gynecol.*, 188(3), 701–707.

Aufderheide, A.C. and Rodriguez-Martin, C. (1998). *The Cambridge Encyclopedia of Human Paleopathology*. Cambridge University Press, Cambridge.

Aveline, C. and Bonnet, F. (2001). Influence de l'anesthésie péridurale sur la durée et les modalités de l'accouchement. *Ann. Fr. Anesth. Reanim*, 20(5), 471–484.

Beauvalet-Boutouyrie, S. (1995). Faut-il supprimer les maternités ? In *L'heureux évènement : une histoire de l'accouchement*. Musée de l'assistance publique-hôpitaux de Paris, Paris.

Beauvalet-Boutouyrie, S. (1999). *Naître à l'hôpital au XIXe siècle*. Belin, Paris.

Berge, C. (1983). Spécificité du petit bassin humain (pelvis minor). Rapport entre filière génitale et volume cérébral. In *Morphologie évolutive, morphogenèse du crâne et origine de l'Homme*, Sakka, M. (ed.). CNRS Éditions, Paris, 115–124.

Berge, C. (1988). Effet de taille et adaptation à la locomotion terrestre chez les primates : analyse multidimensionnelle du pelvis (Primates Catarrhiniens, Carnivores). Implications dans l'évolution des Hominidés. PhD Thesis, Université Paris VII, Paris.

Berge, C. (1991). Quelle est la signification fonctionnelle du pelvis très large de *Australopithecus afarensis* (AL 288-1) ? In *Origine(s) de la bipédie chez les hominidés*, Coppens, Y. and Senut, B. (eds). CNRS Éditions, Paris, 113–119.

Berge, C. (1993). *L'évolution de la hanche et du pelvis des hominidés. Bipédie, parturition, croissance, allométrie*. CNRS Éditions, Paris.

Berge, C. (2003). L'évolution du bassin humain : approche fonctionnelle. In *Anthropologie biologique. Évolution et biologie humaine*, Suzanne, C., Rebato, E., Chiarelli, B. (eds). De Broeck & Larcier, Brussels, 225–231.

Berge, C. and Ponge, J.F. (1983). Les caractéristiques du bassin des australopithèques (A. robustus, A. africanus et A. afarensis) sont-elles liées à une bipédie de type humain ? *Bull. Mém. Soc. Anthropol. Paris*, 10(3), 335–354.

Berge, C., Segebarth-Orban, R., Schmid, P.P. (1984). Obstetrical interpretation of the australopithecine pelvic cavity. *J. Hum. Evol.*, 13, 573–587.

Betti, L. (2017). Human variation in pelvic shape and the effects of climate and past population history. *Anat. Rec.*, 300(4), 687–697.

Boerma, T., Ronsmans, C., Melesse, D.Y., Barros, A.J., Barros, F.C., Juan, L., Moller, A.B., Say, L., Hosseinpoor, A.R., Yi, M. et al. (2018). Global epidemiology of use of and disparities in caesarean sections. *Lancet*, 392(10155), 1341–1348.

Bonnet-Cadilhac, C. (2005). Les accouchements difficiles. In *Maternité et petite enfance en Gaule romaine*, Gourevitch, D., Moirin, A., Rouquet, N. (eds). CEDARC, Bourges, 54–58.

Bouhallier, J. (2000). Analyse 3D de la cavité pelvienne des hominidés : application à l'étude de la diversité de forme des bassins féminins. Master's Thesis, Muséum national d'Histoire naturelle, Paris.

Bouhallier, J. (2006). Évolution de la fonction obstétricale chez les hominoïdes : analyse morphométrique tridimensionnelle de la cavité pelvienne chez les espèces actuelles et fossiles. PhD Thesis, Muséum national d'Histoire naturelle, Paris.

Bouhallier, J. (2015). Le bassin osseux : splendeurs et misères de la clé de voûte du corps humain. In *Mon corps a-t-il un sexe ? Sur le genre, dialogues entre biologies et sciences sociales*, Peyre, E. and Wiels, J. (eds). La Découverte, Paris, 120–141.

Bouhallier, J. (2021). Les accouchements à problèmes à Paris, d'après les archives de la collection Varnier (1874–1903). In *La naissance au risque de la mort. D'hier à aujourd'hui*, Morel M.-F. (ed.). Érès, Toulouse, 171–182.

Bouhallier, J. and Berge, C. (2006). Analyse morphologique et fonctionnelle du pelvis des primates Catarrhiniens : conséquences pour l'obstétrique. *C. R. Acad. Sci.*, 5, 551–560.

Bouhallier, J., Berge, C., Penin, X. (2004). Analyse procuste de la cavité pelvienne des australopithèques (AL 288, Sts 14), des humains et des chimpanzés : conséquences obstétricales. *C. R. Palevol,* 3, 295–304.

Bouvier-Colle, M.H., Ouedraogo, C., Dumont, A., Vangeenderhuysen, C., Salanave, B., Decam, C. (2001). Maternal mortality in West Africa. Rates, causes and substandard care from a prospective survey. *Acta. Obstet. Gynecol. Scand.*, 80, 113–119.

Bowden, D., Winter, P., Ploog, D. (1967). Pregnancy and delivery behavior in the squirrel monkey (*Saimiri sciureus*) and other primates. *Folia Primatol.*, 5(1), 1–42.

Brabin, L. and Brabin, B.J. (1992). The cost of successful adolescent growth and development in girls in relation to iron and vitamin A status. *Am. J. Clin. Nutr.*, 55, 955–958.

Bracken, J.N., Reyes, M., Gendron, J.M., Pierce, L.M., Runge, V.M., Kuehl, T.J. (2011). Alterations in pelvic floor muscles and pelvic organ support by pregnancy and vaginal delivery in squirrel monkeys. *Int. Urogynecol. J.*, 22(9), 1109–1116.

Bruzek, J. (1985). Dimorphisme sexuel de l'os coxal de l'homme du point de vue ontogénique et phylogénique. PhD Thesis, Université Charles, Prague.

Bwana, V.M., Rumisha, S.F., Mremi, I.R., Lyimo, E.P., Mboera, L.E.G. (2019). Patterns and causes of hospital maternal mortality in Tanzania: A 10-year retrospective analysis. *PLoS One*, 14(4), e0214807.

Caldwell, W.E. and Moloy, H.C. (1933). Anatomical variations in the female pelvis and their affect in a labor with a suggested classification. *Am. J. Obstet. Gynecol.*, 26, 479–492.

Cho, F., Hanari, K., Suzuki, M.T., Honjo, S. (1985). Relationship between fetal position and stillbirth in the *Cynomolgus* monkeys (*Macaca fascicularis*): Retrospective analysis. *J. Med. Primatol.*, 14, 169–174.

Chong, M., Pejchinovska, M., Alexander, M. (2024). Estimating causes of maternal death in data-sparse contexts. *Stat. Med.*, 43(24), 4702–4735.

Coates, K., Gibson, S., Williams, L., Kuehl, T.J. (1996). The squirrel monkey as an animal model of pelvic relaxation: An evaluation of a large breeding colony. *Am. J. Obstet. Gynecol.*, 173(6), 1664–1669.

Dastugue, J. and Gervais, V. (1992). *Paléopathologie du squelette humain*. Boubée, Paris.

De Bernis, L. (2005). La mortalité maternelle dans le monde. *Extrait des mises à jour en gynécologie et obstétrique*, XXIX, 93–132.

Diniz, F., Edgington-Giordano, F., Ngo, N.Y.N., Caspi, G., El-Dahr, S.S., Tortelote, G.G. (2024). Morphometric analysis of the intergenerational effects of protein restriction on nephron endowment in mice. *Heliyon*, 10(20), e39552.

Divers, S.J. and Scott, S.J. (2018). *Mader's Reptile and Amphibian Medicine and Surgery*, 3rd edition. Saunders, St. Louis, MO.

Douglas, O. (1983). Mortality of pregnant females in Arikara villages: Osteological evidence. *Amer. J. Phys. Anthropol.,* 61(3), 331–316.

Dumont, A., de Bernis, L., Bouvier-Colle, M.H., Bréart, G. (2001). Caesarean section rate for maternal indication in sub-Saharan Africa: A systematic review. *Lancet*, 358, 1328–1333. doi: 10.1016/S0140-6736(01)06414-5.

Dunworth, H.M. (2021). There is no evolutionary "Obstetrical Dilemma". In *The Routledge Handbook of Anthropology and Reproduction*, Tomori, C. and Han, S. (eds). Taylor and Francis, London.

Falk, D. (1990). Brain size and evolutionary significance of *Homo Sapiens*. *Hum. Evol.*, 5(1), 22–29.

Farabeuf, L.H. and Varnier, H. (1904). *Introduction à l'étude clinique et à la pratique des accouchements*. Steinel, Paris.

Farber, D.M., Giussani, D.A., Jenkins, S.L., Mecenas, C.A., Winter, J.A., Wentworth, R.A., Nathanielsz, P.W. (1997). Timing of the switch from myometrial contractures to contractions in late-gestation pregnant rhesus monkeys as recorded by myometrial electromyogram during spontaneous term and androstenedione-induced labor. *Biol. Reprod.*, 56(2), 557–162.

Farnet-Lepiller, C. (1993). L'enfantement, des origines à la fin de l'Antiquité, en Mésopotamie, en Egypte, en Grèce et à Rome. Thesis, Université Claude Bernard, Lyon.

Fischer, B. and Mitteroecker, P. (2015). Covariation between human pelvis shape, stature, and head size alleviates the obstetric dilemma. *Proc. Natl. Acad. Sci. USA*, 112, 5655–5660.

Gelis, J. (1995). De la matrone à la sage-femme et à l'accoucheur ou les mutations de l'accouchement. In *L'heureux évènement : une histoire de l'accouchement. Catalogue d'exposition*. Musée de l'assistance publique-hôpitaux de Paris, Paris.

González, A.M. and Chamero, P. (2017). Biomechanical constraints and obstetric complications: The influence of the obstetric dilemma on human birth. *J. Evol. Biol.*, 30, 63–76.

Hausler, S. (2021). The challenges of human birth: Obstetric dilemmas and pelvic anatomy. *Am. J. Phys. Anthropol.*, 172(1), 1–11.

Henrion, R., Delecour, M., De Tourris, H. (1994). *Gynécologie et obstétrique*, 6th edition. Masson, Paris.

Hossain, A.T., Siddique, A.B., Jabeen, S., Khan, S., Haider, M.M., Ameen, S., Tahsina, T., Chakraborty, N., Nahar, Q., Jamil, K. et al. (2023). Maternal mortality in Bangladesh: Who, when, why, and where? A national survey-based analysis. *J. Glob. Health.*, 9(13), 07002.

Joulin, D. (1864). Anatomie et physiologie comparée du bassin des mammifères. *Archives générales de médecine*, 29–46.

Kawada, M., Nakatsukasa, M., Nishimura, T., Kaneko, A., Morimoto, N. (2020). Covariation of fetal skull and maternal pelvis during the perinatal period in rhesus macaques and evolution of childbirth in primates. *Proc. Natl. Acad. Sci. USA*, 117(35), 21251–21257.

Kawada, M., Nakatsukasa, M., Nishimura, T., Kaneko A., Ogihara, N., Yamada, S., Coudyzer, W., Zollikofer, C.P.E., Ponce de León, M.S., Morimoto, N. (2022). Human shoulder development is adapted to obstetrical constraints. *Proc. Natl. Acad. Sci. USA*, 119(16), e2114935119.

Keith, A. (1924). On the origin and nature of hernia. *Br. J. Surg.*, 11(43), 455–475.

Kissler, K. and Hurt, K.J. (2023). The pathophysiology of labor dystocia: Theme with variations. *Reprod. Sci.*, 30, 729–742.

Krogman, W. (1951). The scars of human evolution. *Sci. Am.*, 185, 54–57.

Kurki, H. (2007). Protection of obstetric dimensions in a small-bodied human sample. *Am. J. Phys. Anthropol.*, 133, 1152–1165.

Kurki, H. (2023). The evolution of the human birth canal: A new approach to understanding obstetric difficulties. *Curr. Anthropol.*, 64(3), 355–370.

Le Damany, G. (1912). L'adaptation de l'Homme. *J. Anat. Physiol.*, XII, 169.

Leutenegger, W. (1974). Functional aspects of pelvic morphology in simian primates. *J. Hum. Evol.*, 3, 207–222.

Leutenegger, W. (1982). Encephalization and obstetrics in primates with particular reference to human evolution. In *Primate Brain Evolution: Methods and Concepts*, Armstrong, E. and Falk, D. (eds). Plenum Press, New York/London, 85–95.

Lieverse, A.R., Bazaliiskii, V.I., Weber, A.W. (2015). Death by twins: A remarkable case of dystocic childbirth in Early Neolithic Siberia. *Antiquity*, 89(343), 23–38.

Lindburg, D.G. (1982). Primate obstetrics: The biology of birth. *Am. J. Primat.*, 3(1), 193–199.

Maestripieri, D. and Mateo, J. (2009). *Maternal of Stress Effects in Mammals*. Chicago University Press, Chicago.

Mafart, B.Y. (1994). Approche de la mortalité naturelle au Moyen Âge en Provence. *Dossiers de documentation archéologique, CNRS*, 17, 207–219.

Malinas, Y. and Favier, M. (1979). *Mécanique obstétricale*. Masson, Paris.

Manfredini, M. (2020). The effects of nutrition on maternal mortality: Evidence from 19th–20th century Italy. *J. Obstet. Gynaecol.*, 32(6), 787–793.

Maygrier, C. (1893). Le service d'accouchements de "la Pitié" depuis 1886. *Ann. Soc. Obstet. Fr.*, 1(2/2), 8.

McCue, M. and Ferris, R.A. (2012). Parturition, dystocia and foal survival: A retrospective study of 1047 births. *Equine Veterinary J.*, 41, 22–25.

Mitteroecker, P., Hutteggerb, S.M., Fischer, B., Pavlicev, M. (2016). Cliff-edge model of obstetric selection in humans. *Proc. Natl. Acad. Sci. USA*, 113(51), 14680–14685.

Papillault, R. (1902). L'homme moyen à Paris. Variation suivant le sexe et suivant la taille. Recherches anthropométriques sur 200 cadavres. *Bull. Mém. Soc. Anthropol. Paris*, 3, 393–526.

Pavličev, M., Romero, R., Mitteroecker, P. (2020). Evolution of the human pelvis and obstructed labor: New explanations of an old obstetrical dilemma. *Am. J. Obstet. Gynecol.*, 222(1), 3–16.

Pereira da Silva, G., Pereira de Souza, T.H., Lima, F., Kelen, A., Vicente, R., Ricardo, W., Kuehl, T.J., Ruiz, J.C., Monteiro, F.O.B. (2021). Female squirrel monkeys as models for research on women's pelvic floor disorders. *Lab. Anim.*, 55(6), 499–508.

Pinka, K., Fischer, B., Huffmane, M.A., Miyabe-Nishiwakif, T., Suda-Hashimotof, N., Kanekof, A., Wallnerg, B., Pflügerg, L.S. (2024). No birth-associated maternal mortality in Japanese macaques (*Macaca fuscata*) despite giving birth to large-headed neonates. *Proc. Natl. Acad. Sci. USA*, 121(42), 1–8.

Ponce de León, M., Golovanova, L., Doronichev, V., Romanova, G., Akazawa, T., Kondo, O., Ishida, H., Zollikofer, C.P.E. (2008). Neanderthal brain size at birth provides insights into the evolution of human life history. *Proc. Natl. Acad. Sci. USA*, 105(37), 13764–13768.

Quereux, C., Gabriel, L., Schaal, J.-P. (1995). *Accouchement normal en présentation du sommet : mécanisme physiologique et conduite à tenir*. Ellipses/AUPEF/UREF, Paris.

Rasmussen, K., Ausman, L.M., Hayes, K.C. (1980). Vital statistics from a laboratory breeding colony of squirrel monkeys (*Saimiri sciureus*). *Lab. Animal. Sci.*, 30(1), 99–106.

Reynolds, E. (1931). The evolution of the human pelvis in relation to the mechanics of the erect posture. *Papers of the Peabody Museum of American Archaeology and Ethnology University*, 11(5), 255–334.

Ridley, M. (1995). Brief communication: Pelvic sexual dimorphism and relative neonatal brain size really are related. *Am. J. Phys. Anthropol.*, 97(2), 197–200.

Rosenberg, K.R. (1992). The evolution of modern human childbirth. *Yearbook of Physical Anthropology*, 35, 89–124.

Rosenberg, K.R. and Trevathan, W.R. (2000). The shoulders follow the head: Post-cranial constraints on human childbirth. *J. Hum. Evol.*, 39, 583–586.

Rosenberg, K.R. and Trevathan, W.R. (2005). Bipedalism and human birth: The obstetrical dilemma revisited. *Evol. Anthropol.*, 4(5), 161–168.

Rosenberg, K.R. and Trevathan, W.R. (2007). An anthropological perspective on the evolutionary context of preeclampsia in humans. *J. Reprod. Immunol.*, 76(1/2), 91–97.

Rosenberg, K.R. and Trevathan, W.R. (2018). Evolutionary perspectives on cesarean section. *Evol., Med., Public Health*, 1, 67–81.

Ruff, C.B. (1994). Morphological adaptation to climate in modern and fossil humans. *Yearbook of Physical Anthropology*, 37, 65–107.

Ruff, C.B. (2017). Mechanical constraints on the hominin pelvis and the "obstetrical dilemma". *The Anatomical Record*, 300, 946–955.

Saucedo, M. and Deneux-Tharaux, C. (2024). Mortalité maternelle en France, 2016–2018, fréquence, causes et profil des femmes. *Gyn. Obstet. Fertil. Senol.*, 52(4), 185–200.

Say, L., Chou, D., Gemmill, A., Tunçalp, O., Moller, A.B., Daniels, J., Metin Gülmezoglu, A., Temmerman, M., Alkema, L. (2014). Global causes of maternal death: A WHO systematic analysis. *Lancet Glob. Health.*, 2(6), e323–33.

Schlumbohm, J. (2002). Comment l'obstétrique est devenue une science. La maternité de l'Université de Göttingen. *Actes de la recherche en sciences sociales*, 143, 18–30.

Schofield, R. (1986). Did the mothers really die? Three centuries of maternal mortality. In *The World We Have Gained. Histories of Population and Social Structure*, Bonfield, L., Smith, R.M., Wrightson, K. (eds). Basil Blackwell, Oxford/New York, 231–260.

Schultz, A.H. (1949). Sex differences in the pelves of primates. *Am. J. Phys. Anthropol.*, 7, 401–423.

Schultz, A.H. (1961). Vertebral column and thorax. In *Primatologia. Handbook of Primatology*, Hofer, H., Schultz, A.H., Starck, D. (eds). Karger, Basel/New York.

Sibley, L.M., Armelagos, G.J., Van Gerven, D.P. (1992). Obstetric dimensions of the true pelvis in a medieval population from Sudanese Nubia. *Am. J. Phys. Anthropol.*, 89, 421–430.

Snow, W. (1952). *Roentgenography in Obstetrics and Gynecology*. Charles C. Thomas, Springfield.

Starrach, T., Schmidhuber, L., Elger, L., Franz, M., Buechel, J., Hübener, C., Kolben, T., Koliogiannis, V., Mahner, S., Hasbargen, U. et al. (2022). Pelvic inlet area is associated with birth mode. *Acta Obstet. Gynecol. Scand.*, 102(1), 59–66.

Tague, R.G. (1991). Commodities in dimorphism and variability in the anthropoid pelvis, with implications for the fossil record. *J. Hum. Evol.*, 21, 153–176.

Tague, R.G. (1994). Maternal mortality or prolonged growth: Age at death and pelvic size in three prehistoric Amerindian populations. *A.J.P.A.*, 95, 27–40.

Tague, R.G. and Lovejoy, C.O. (1986). The obstetric pelvis of AL 288-1 (Lucy). *J. Hum. Evol.*, 15, 237–255.

Tarnier, S. (1858). *De la fièvre puerpérale observée à l'hospice de la Maternité*. Baillère, Paris.

Taub, D.M., Adams, M.R., Auerbach, K.G. (1978). Reproductive performance in a breeding colony of Brazilian squirrel monkeys (Saimiri sciureus). *Lab. Anim. Sci.*, 28(5), 562–566.

Thoms, H. (1941). Clinical application of roentgen pelvimetry and study of results in 1100 white women. *Am. J. Obst. Gynec.*, 42(6), 957–975.

Topinard, P. (1990). *L'Homme dans la nature*. Éditions Jean-Michel Place, Paris [Original published 1891].

Touraille, P. and Gouyon, P.H. (2008). Why are women smaller than men? When anthropology meets evolutionary biology. *Nat. Preced.*, 3. doi: 10.1038/npre.2008.1832.1.

Trevathan, W.R. (1988). Fetal emergence patterns in evolutionary perspective. *Am. Anthropol.*, 90, 674–681

Trevathan, W.R. (1996). The evolution of bipedalism and assisted birth. *Med. Anthropol. Quaterly*, 10(2), 287–298.

Ullrich, H. (1975). Estimation of fertility by means of pregnancy and childbirth alterations at the pubis, the ilium, and the sacrum. *OSSA*, 2(1), 23–39.

Vilos, G.A., Vilos, A.G., Burbank, F. (2024). Bipedalism and the dawn of uterine fibroids. *Hum. Reprod.*, 39(3), 454–463.

Wall-Scheffler, C.M. and Kurki, H. (2022). Changing human birth patterns: New insights into the obstetric dilemma and evolutionary adaptations. *PLOS One*, 17(10), e0268260.

Walrath, D.E. (2003). Rethinking pelvic typologies and the human birth mechanism. *Curr. Anthropol.*, 44(1), 5–31.

Waltenberger, L., Rebay-Salisbury, K., Mitteroecker, P. (2022). Age dependent changes in pelvic shape during adulthood. *Anthropol. Anz.*, 79(2), 143–156.

Warrener, A.G., Lewton, K., Pontzer, H., Lieberman, D.E. (2015). A wider pelvis does not increase locomotor cost in humans, with implications for the evolution of childbirth. *PLOS One*, 10(3), e0118903.

Washburn, S.L. (1960). Tools and human evolution. *Sci. Am.*, 203(3), 63–72.

Waterman, H.C. (1929). Studies on the evolution of the pelvis of man and other primates. *Bull. AMNH*, 58, 12.

Weaver, T.D. (2002). A multi-causal functional analysis of hominid hip morphology. PhD Thesis, Stanford University, Stanford.

Webb, N., Fornai, C., Krenn, V.A., Watson, L.M., Herbst, E.C., Haeusler, M. (2024). Gradual exacerbation of obstetric constraints during hominoid evolution implied by re-evaluation of cephalopelvic fit in chimpanzees. *Nat. Ecol. Evol.*, 8, 2228–2238.

Weiner, S., Monge, P., Mann, G. (2008a). Cesarean sections and the evolutionary cost of human birth. *J. Evol. Biol.*, 25(12), 1519–1528.

Weiner, S., Monge, J., Mann, A. (2008b). Bipedalism and parturition: An evolutionary imperative for cesarean delivery? *Clin. Perinatol.*, 35(3), 469–478.

Wells, J.C.K. (2015). Between Scylla and Charybdis: Renegotiating the resolution of the "obstetric dilemma" in response to ecological change. *Philos. Trans. R. Soc. B. Biol. Sci.*, 370(1666), 20140067.

Wells, J.C.K. (2017). The new "obstetrical dilemma": Stunting, obesity and the risk of obstructed labour. *Anat. Rec.*, 300, 716–731.

Wells, J.C.K., DeSilva, J.M., Stock, J.T. (2012). The obstetric dilemma: An ancient game of Russian roulette, or a variable dilemma sensitive to ecology? *Am. J. Phys. Anthropol.*, 55, 40–71.

Wells, J.C.K., Wibaek, R., Poullas, M. (2018). The dual burden of malnutrition increases the risk of cesarean delivery: Evidence from India. *Front. Public Health*, 6, 292.

Wheeler, P.E. (1991). The influence of bipedalism on the energy and water budgets of early hominids. *J. Hum. Evol.*, 21(2), 117–136.

Witcher, R. (2019). Editorial. *Antiquity*, 93(368), 287–296.

7

How Upright Posture Endures Despite Deficits in Core Postural Motor Control

7.1. Introduction

Examining the coordination between posture and movement is a fundamental element of clinical investigation in children with motor disabilities. Parents of a young child with delayed motor development often seek the PRM (Physical and Rehabilitation Medicine) physician's authority to provide insight into prognosis; rehabilitation therapists look for guidance to shape a course of action; and the child, implicitly, hopes for any intervention that may support development.

Yet, despite decades of practice across neurology, the collective experience of teams and families, the lessons passed down by our teachers, and the study of manuals, articles and best-practice guides, we still have few reliable tools to ensure that our medical judgment reflects the child's true trajectory. When the rare clinical or paraclinical data available for motor prognosis are challenged, the fragility of our conclusions becomes evident.

Paradoxically, these moments of uncertainty are essential. They compel us to re-examine developmental hypotheses and to search more deeply for the resources through which each child may progress. In this way, the very confusion that unsettles us also drives innovation and discovery.

Chapter written by Véronique LEROY-MALHERBE, Élise PINEAU and Virginie PINEAU.

7.2. Report of two clinical cases

7.2.1. *First clinical case: Jeremy, affected with cerebral palsy*

Jeremy is a 16-year-old boy with cerebral palsy who is able to walk outdoors. After acute fetal distress, he was monitored in intensive care for one week during the perinatal period, followed by two weeks in the pediatric unit. His initial brain MRI was normal. At hospital discharge, the pediatrician noted abnormal tone.

By age 2, his motor impairment was defined as spastic diplegia with right-sided predominance, accompanied by marked dyskinetic movements. Motor development was delayed, but he learned to crawl, move on all fours and attempt sitting. He was attentive and motivated in motor exercises and daily mobility. Orofacial involvement led to speech disorders without swallowing difficulties. His articulation was severely impaired, rendering speech incomprehensible to unfamiliar listeners, although language development was appropriate. Cognitive abilities and visuospatial construction were strong.

At age 4, the contrast between his determination to exercise motor skills and his significant postural disorder was striking. He exhibited cervical extension (cervid-like posture) and passive thoracic kyphosis when seated, suggesting a persistent postural tone disorder. The apparent "hump" in sitting, which disappeared with modified supports, indicated a neurological rather than structural origin. Jeremy reacted strongly to external stimuli: sudden noises or tactile input caused him to freeze with cervical extension and arms flexed in an elevated "chandelier-like" posture. Spasticity was troublesome on the right side, with equinus positioning of the foot when upright. Despite this, selective motor control was excellent for his age, and no orthopedic consequences were noted.

Rehabilitation followed developmental motor stages (Le Métayer 1999), emphasizing awareness of trunk extension to the apex of the head. He used a walker indoors and was assisted at home by adults supporting his shoulders to enhance ground reaction forces and encourage trunk straightening. He otherwise moved on all fours. A manual wheelchair was not prescribed, partly due to parental reluctance and partly to preserve emerging corrections in sitting posture.

At age 5, Jeremy unexpectedly managed 10 independent indoor steps. Shoes with wide soles were prescribed to provide a broad base of support. With this aid, trunk straightening improved, and cervical lordosis was voluntarily corrected. Encouragement, visual fixation or tactile stimulation along the spine further enhanced posture. Speech production also improved, becoming comprehensible to all.

At age 6, botulinum toxin injections were administered to bilateral hamstrings, triceps, tibialis posterior and tibialis anterior muscles. Within three weeks, right-sided equinus resolved, enabling stable standing. Indoors, he relaxed, turned and maintained endurance with a rollator outdoors, although he remained unable to stop without support. Two months later, he could stand unaided and take several consecutive steps, using his arms in a pendular rather than elevated posture. Squat-to-stand transitions were well established, and startle reactions no longer triggered abnormal limb positioning.

At age 6 and a half, neurological examination again revealed massive postural tone disorder, with trunk collapse upon destabilization in vertical, seated or lateral suspension positions. Rehabilitation focused on trunk and head control in sitting and standing. Jeremy cooperated with this demanding program, which yielded clear improvement by age 10.

At age 10, he walked independently outdoors, adapted to varied terrain and stopped without losing balance. Speech was intelligible, and breathing regulation reflected improved control of dyskinetic orofacial movements. However, gait analysis revealed pelvic jerking with lateral shifts during weight-bearing, although spinal alignment remained intact in the frontal plane, with lumbar hyperlordosis and stable head control. Pelvic radiography confirmed orthopedic stability, with hips well positioned. To conserve energy and minimize pelvic jerks, he was advised to use a walker.

At age 16, Jeremy's ambulation and motor skills were similar to those at 10. Orthopedic status remained stable, but neurological examination continued to show the same original deficits in postural regulation.

7.2.2. *Second clinical case: Lisa, affected with Friedreich's ataxia with posterior cord bundle dysfunction*

Lisa, aged 8, presents with a two-year history of progressive sensory neuropathy leading to loss of independent walking. This corresponds to a rapidly progressive form of Friedreich's ataxia, without cerebellar signs (no hypotonia, dysarthria, abnormal tendon tone or oculomotor disorder) and without pyramidal signs (no Babinski response). Clinical examination indicates deep sensory impairment: dynamic ataxia compensated by visual control, reduced vibration sense and absent osteotendinous reflexes. Electromyography shows isolated proprioceptive impairment, with complete absence of sensory conduction responses and preserved motor activity, suggesting likely involvement of Clarke's column in the posterior horn of the spinal cord. Genetic testing confirms Friedreich's ataxia with compound heterozygosity.

Lisa is lively and playful, eager to participate in adapted motor activities. She can no longer stand unaided but moves forward on all fours, supporting herself on her limbs. When held at the chest or under the armpits, she alternates her lower limbs with small thigh elevation and regular stepping. Foot support is stable, without deformity. At the end of the stance, knee hyperextension stabilizes the joint, allowing cautious trunk advancement over the center of pressure. She can straighten her arms and use her walker to pull herself upright, walking indoors at a steady pace, turning, stopping on request and reaching objects at her height. Sitting, she can catch a ball, but her trunk destabilizes quickly. From squat to stand, she maintains vertical alignment if her eyes are open and knees locked.

Proprioception testing highlights deficits: with eyes closed and no auditory or vestibular cues, she cannot reproduce limb positions after passive placement, confuses upper and lower limbs and fails to detect initiation of movement in any segment. Muscle strength is preserved (MRC scale), demonstrated by her ability to support body weight in a "wheelbarrow" position. She moves on all fours with a stable pelvis and coordinated sequences, resisting imposed movements.

There are no pyramidal or cerebellar signs. Visual control allows accurate finger-to-target movements. Motor organization remains preserved, enabling coordinated progression on all fours and with a walker, adapting to narrow spaces and maintaining pace. She independently transfers into her wheelchair, using a sequential, anticipatory strategy: pulling up from all fours, kneeling, straddling and rotating onto the seat. Movements are harmonious, precise and fatigue-free.

Lisa develops a 60° kyphoscoliosis, flexible and manually reducible, requiring osteosynthesis surgery. Postoperatively, spinal alignment is restored along the osteosynthesis, but cervical anterior translation and sacral horizontalization with hip flexion contracture are noted.

Over two years of follow-up, she continues to perform unassisted transfers and maintains her motivation to move, supported by physiotherapy.

7.3. Discussion

7.3.1. *Two fundamental neurological systems in walking*

Two neurological systems are fundamental to walking:

– The first governs upright stance and step alternation, mediated by the monosynaptic spinal reflex arc and the spinal gait generator.

– The second regulates posture and step progression, relying on proprioceptive spinal pathways.

7.3.1.1. *A system for standing upright or alternating steps*

Bipedal walking involves both postural activity and movement activity.

– The postural control system is based on the monosynaptic stretch reflex, which maintains upright stance.

– The locomotor system is driven by the spinal gait generator, producing rhythmic alternation of the lower limbs in response to plantar stimulation.

The physiological basis of posture through the monosynaptic arc can be summarized as follows:

– The muscle is stretched.

– The neuromuscular spindle transmits this information to the motor neuron.

– The motor neuron triggers reflex contraction of the same muscle.

This reflex generates the fundamental tone required to stand upright and to readjust any deviation from vertical alignment. For example, in a standing subject, forward destabilization stretches the posterior muscle chain, eliciting reflex contraction and restoring verticality. A similar mechanism occurs in the anterior chain during backward destabilization. One hypothesis is that this stabilization system emerged through evolution when the need to sustain bipedal posture arose 7–8 million years ago, alongside skeletal adaptations (Tardieu 2018) and natural selection.

In the paleoanthropological scenario known as the East Side Story (Coppens 2000), humans moved from forest to savanna. The need to hunt and scan the horizon over tall grass replaced the need to climb trees for fruit, favoring upright posture. The stretch reflex can thus be understood as nature's way of harnessing gravity to maintain stance without conscious effort. Being monosynaptic, it is simple; being spinal, it is automatic; and by producing reflex contraction during muscle stretch, it generates tone throughout the body without cognitive input.

The rhythmic alternation of the limbs is generated and maintained by the central gait generator, which relies on polysynaptic reciprocal-inhibition circuits in the spinal cord. These circuits, present in vertebrates and invertebrates, regulate rhythmic patterns and limb alternation in both quadrupeds and bipeds. At slow or moderate speeds, direct reciprocal inhibition between homolateral and contralateral muscles enables step alternation. At faster speeds, indirect contralateral inhibition via excitatory interneurons connected to inhibitory interneurons amplifies the signal, allowing running with alternating steps rather than leaping (Bouvier 2014).

Both systems – postural control and limb alternation – are active from birth. Automatic walking in the newborn is triggered by holding the infant upright with the trunk slightly inclined forward and stimulating the sole, producing regular alternating steps (Kiehn 2011). The abduction reaction of the lower limb during lateral suspension (Vasseur 2000) demonstrates efficient pyramidal pathway activity against gravity, even though in-utero gravitational experience is limited. In infant neurological assessment, the quality of these gravity reactions helps detect pyramidal sequelae. Walking, as a succession of imbalances (single support) and balance recoveries (double support), depends primarily on automatic trunk stabilization.

Biomechanical studies confirm that trunk alignment during mid-stance is fundamental (MacKinnon et al. 1993). The pelvis, particularly the hip, acts as the pivot for trunk balance. Clinical evaluation of walking ability focuses on this region (Leroy-Malherbe 2017a). A persistent collapse in lateral suspension, or inability to maintain sitting posture at age 2, are reliable negative prognostic markers for independent ambulation in cerebral palsy (Sala et al. 1995; Truscelli 2017).

In Jeremy's case, all clinical evidence suggested that independent walking was unattainable. His neurological circuitry failed to establish the necessary automatic reactivity to gravity, with consequences in both standing and sitting. Notable signs included thoracic kyphosis and deer-like cervical posture persisting beyond age 7 and non-reactive lateral suspensions. Yet, Jeremy eventually acquired an automatic gait, defined as ambulation of more than three unaided successive steps, with anticipatory regulation of vertical acceleration of the center of gravity across the entire sequence rather than step-by-step – a phenomenon previously observed in hemiplegic children (Leroy-Malherbe et al. 1992).

7.3.1.2. *A system to regulate posture or the advancement of steps*

Posture and the alternation of limb movements are enabled by muscular contractions resulting from reflex circuits that keep the body upright or propel it forward. Yet it is proprioception that integrates the joints with one another, ensuring fine regulation and rapid adaptation to environmental demands (Leroy-Malherbe 2018).

The neuromuscular spindle is the organelle that detects changes in muscle length and the speed of stretching. It is the primary site of proprioceptive signaling, which builds awareness of muscle tension, joint positioning and the representation of the body in motion. This small organelle, located on specific muscle fibers near the tendon insertion, can be compared to a spring encircling the muscle fiber. It enables the muscle to react to changes in length, stretching velocity and mechanical vibration applied to the tendon. The spindle provides information about joint position, detects movement and calculates the strength of muscular tension.

Proprioception regulates balance, controls movement and stabilizes the joints. It is essential for standing, for positioning the body in space under conditions of balance or imbalance and for mentally projecting oneself into action. Sustained stimulation of proprioception through tendon vibration generates a representation of the body in motion: the vibrated muscle is perceived "as if elongated." For example, vibration of the quadriceps tendon induces a sensation of knee flexion, which can facilitate trunk straightening (Roll et al. 1986). This effect has also been observed in individuals with cerebral palsy, who can straighten from a collapsed sitting position when such stimulation is applied (Redon-Zouiteni et al. 1994). In typically developing children, proprioception contributes to the correction of future motor acts, playing a key role in motor learning between the ages of 5 and 8, and continuing through adolescence up to around age 24 (Wallard et al. 2016).

Clinically, measuring proprioceptive deficits is challenging in adults and nearly impossible in young children. Testing requires minimizing the contribution of other sensory systems involved in balance, particularly vision and the vestibular apparatus. Three types of tests are commonly used (Bruyneel 2022):

– Position matching: measuring the difference between a limb-segment position held for three seconds and the "assumed identical" position reproduced without the target. This can be assessed via passive mobilization, active reproduction guided by an image or an instructed active movement.

– Movement detection: recording the angle or time at which the subject detects a movement induced manually or mechanically.

– Force perception: detecting the direction of an applied force or reproducing a previously imposed force on a dynamometer.

Neurology shows that impairment of proprioceptive afferents is particularly disruptive for vertical orientation. It leads to difficulties in constructing the body axis, the body schema and fine motor skills during critical developmental periods in childhood and adolescence. The body schema is the internalized representation of the "situated body" in space (Assaiante et al. 2014). It expresses how the body maintains itself in the face of gravity and moves through space.

Amblard demonstrated that in microgravity, and in the absence of vision, destabilization of the trunk is compensated by head movements. These compensations cannot rely on proprioceptive, visual or vestibular receptors but instead require reference to a memorized representation of the body. During navigation under microgravity, head movements precisely counterbalance trunk movements (Amblard et al. 1997). This highlights the existence of a body representation based on the geometry of the body itself, rather than on sensory feedback.

In the first clinical case, Jeremy appears to rely predominantly on this internalized representation. In the second case, Lisa loses her body references as soon as she closes her eyes. Her body representations may lack a structured basis, particularly those not yet consolidated due to her young age, such as laterality. Yet she manages to maintain postural control with a walker and systematically coordinates gross motor skills to transition between positions. Her ability to remain upright despite a 60° scoliosis, and her rapid recovery of axial alignment after corrective surgery, suggest that her perception of her center of gravity is independent of her trunk's perceived position. Instead, it likely derives from her visual perception of the orthonormal coordinates of her body in space.

7.3.2. *What resource factors despite these two fundamental neurological deficits?*

We propose that two non-cognitive factors contribute to the compensatory mechanisms observed in these children:

– biomechanical factors, which render walking an automatic activity requiring minimal muscular effort;

– sensory factors, which, beyond proprioception, participate in the regulation of gait.

7.3.2.1. *Biomechanical factors of walking*

Posture and movement, the two functions essential to motor skills, are inseparable from the experience of gravity. Walking is a process in which posture and movement are tightly coordinated, making Earth's gravity an ally rather than a burden. The body reacts in the face of gravity, not against gravity. Through the perception of ground reaction forces and the body's orientation relative to the gravitational vertical, a walking subject optimizes balance and displacement, as demonstrated in studies on gait initiation.

During the initiation of walking, the transition from a predominantly postural activity to a predominantly motor activity occurs, in the able adult, from the very first step (Brenière et al. 1986). From this moment onward, the adult is capable of adjusting the instantaneous velocity of the center of gravity to match the speed of progression (Brenière et al. 1987). In contrast, in a young child who has just learned to walk, regulation of progression speed is not observed until the third or fourth step (Brenière et al. 1989). These initial steps are interpreted as a forward fall, since the child cannot yet manage the biomechanical parameters of the body – mass and moment of inertia – and lacks a motor intention of progression at a determined speed.

After approximately 18 months of independent walking – around the age of 3 – mature gait is acquired. At this stage, the child has experienced and integrated the biomechanical references of the body. The position of the center of gravity relative to the barycenter of the ground's center of pressure is controlled. The sequence of walking phases becomes automated, with a constant instantaneous velocity of the center of gravity from the first step (Bril et al. 1992).

From a global perspective, based on the body's center of gravity, an inverted pendulum model has been proposed to mathematically represent human gait (Winter 1995). In this model, the body is considered fixed at its lower point (the feet) and mobile around the vertical axis at its upper end (the head), with a displacement corresponding to step length and an oscillation frequency corresponding to step cadence. Progression speed is directly determined by step length and frequency.

Analysis of gait parameters – step length, step frequency, progression speed and instantaneous velocity of the center of gravity – aligns with this model and supports the concept of gait as an economical and potentially infinite motor activity, akin to a pendulum swinging indefinitely in a frictionless environment. Within this global model, the pelvis – stabilized by forces applied at the hip – emerges as the regulating factor of locomotor automatism. Indeed, the pelvis is the skeletal segment closest to the body's center of gravity (Yang et al. 2014), which is located at L5–S1 (Duval-Beaupère et al. 1989). Calculations of torques applied at the hip (Brenière et al. 1991; Winter et al. 1993) confirm the efficiency of this biomechanical regulation.

From a segmental perspective of the body, torques, work and power are analyzed at each joint. This model confirms that a significant portion of the effort is required at the hip to stabilize the frontal plane (Eng et al. 1995). Studies emphasize the landing of the swinging step beneath the pelvis as a determinant of energy expenditure in regaining balance. The placement of the step at the end of the swing ensures, at a given moment, correspondence between the center of gravity and the barycenter of the ground's center of pressure (Kuo et al. 1999; Collins et al. 2005; Wang et al. 2014; Bruijn et al. 2018). This principle applies equally to the sagittal and frontal planes, independently of one another.

In the sagittal plane, flexion and extension of the thigh play a major role, enabling coordination between the dynamics of the head, trunk and upper limbs. The sequence of lower-limb movements in the sagittal plane is essentially ballistic, characterized by energy generation and recovery between muscles with eccentric activity and those with concentric activity, or between the proximal and distal portions of biarticular muscles (Elftman 1939). In subjects with cerebral palsy, however, segmental regulation and the resulting energy transfers are disrupted by co-contractions (Winter 1990). These phenomena, initially subclinical in cerebral palsy infants, emerge early in biarticular muscles and impair motor regulation of gait.

In the frontal plane, the segmental model reinforces the principle of muscular economy. By placing the swinging step beneath the pelvis, activity of the foot's invertors/evertors on the swinging limb is minimized, while the supporting hip's abductors are optimized. During single-leg support, lateral imbalance of the trunk, head and arms above the pelvis is largely compensated by passive pelvic stabilization linked to step placement. This placement induces mediolateral acceleration at the hip and subtalar joint of the swinging leg. Such acceleration generates most of the balancing force, reducing the need for active recruitment of hip abductors in both the swinging and supporting limbs (Winter et al. 1993).

Further evidence supports this principle of hip economy in lateral stability. During the swaying phase of the step, rapid latency responses recorded in the gluteus medius of the swinging limb indicate a passive mechanical phenomenon rather than voluntary contraction. The absence of effect from neuromuscular block applied during this phase reinforces the observation that limb displacement is achieved passively through biomechanics (Pohl et al. 2015). Clinically, this has long been observed in patients with complete gluteus medius deficiency following poliomyelitis, who remain able to walk with a cane. Thus, to facilitate lateral stability and step advancement, the body maximizes passive dynamics, selecting lateral step placement that optimizes energy cost, a positioning linked to pelvic dimensions (MacGeer 1990). Muscular effort of the hip abductors is thereby minimized. Active contraction and coordination around the pelvis only occur when navigation strategies change – for example, to correct imbalance (MacKinnon et al. 1993). During these changes in movement patterns, body dynamics may also involve changes in the moments of force exerted by the ankle on the supporting foot, either through propulsion or through evertor/invertor contractions, or they may involve small, rapid tilts of the trunk (Bruijn et al. 2018).

This "low-cost" lateral balance is consistent with studies of walking on the spot based on the global model (Brenière 1998), which highlight a natural body frequency – a universal invariant in humans. Since the amplitude of lateral oscillations of the center of mass during walking on the spot depends on walking frequency, it can be suggested that subjects naturally adjust their gait frequency towards this intrinsic body rhythm, thereby optimizing lateral oscillation amplitude biomechanically.

Whether balance in walking is studied in the frontal or sagittal plane, these studies demonstrate that walking is fundamentally a self-evident human function, with dynamics that optimize the intrinsic properties of body kinetics.

The particularity of the two reported cases lies in the preservation of upright posture where it would not normally be expected. In Jeremy's case, as a young child, he retained segmental control and thus the ability to organize paradoxical movements – flexion at one end of a limb segment and extension at the other. Rehabilitation may have focused his attention on shifts in his center of gravity, encouraging compensatory strategies less economical than passive step placement: ankle strategy, trunk strategy or head strategy. An ankle strategy involves contractions around the ankle; a trunk strategy involves contractions around the pelvis caused by trunk tilt.

As previously said, both are observed in typical subjects when gait dynamics become irregular. A head strategy, characterized by non-lateral anchoring of the head on the trunk, has been noted in cerebral palsy children with "jump gait" (Wallard 2014). Jeremy's trunk hypotonia, step jolts and increased lateral imbalance may not have caused expected difficulties if foot placement adequately compensated for trunk acceleration. Contributing factors may include shoes with large soles, which increase ground support surface, and botulinum toxin treatment, which eliminated his dynamic equinus on the right.

A further hypothesis involves stimulation of plantar tactile receptors: the accentuated, noisy foot landing may have activated them. This rhythmic aspect has been shown to play a role in gait regulation in automated movement in cerebral palsy subjects (Leroy-Malherbe 1992).

For Lisa, who likely lacks an internalized sense of body position in space, compensation appears to rely entirely on vision and deliberate observation of step placement. She demonstrates significant ability to process dual motor tasks – advancing towards a goal, speaking and observing foot placement simultaneously. With these capacities preserved, ballistic lower-limb movement, if correctly initiated through sight, may function passively to allow forward progression. Her weight-bearing step ends with knee recurvatum, which locks the joint and enables her to advance her center of gravity without falling. It must be noted, however, that her ability to stand is only possible with walker support, which reduces lateral imbalance.

7.3.2.2. *Sensory information involved in posture and regulation of walking*

The main neurosensory functions that ensure postural control are peripheral vision and the vestibular system, in addition to muscle proprioception, which has already been discussed.

The visual flow provided by the peripheral retina makes a major contribution to posture regulation (Bullinger et al. 1996; Vasseur 2000). This effect is accentuated by corridors of alternating black-and-white vertical stripes or Vichy panels. Their placement on either side of a subject – whether standing, walking or sitting – affects trunk regulation in the sagittal plane. Such visual flow improves sitting posture, gripping ability and trunk and head straightening during walking, both in healthy subjects and in those with cerebral palsy, particularly at the early stages of compensation (Barale et al. 2016).

The vestibular system provides information on head accelerations and contributes to balance through rapid postural control. Its compensatory role in children with cerebral palsy has not been studied, and its contribution during steady-state walking is minimal. Berthoz demonstrated that during upright progression, the orbitomeatal axis functions as a stable reference point for the body (Berthoz 2013). This head-based reference emerges early in walking development, approximately 10–15 weeks after independent walking begins (Ledebt et al. 1995). Head stabilization thus appears to be a primary mechanism of postural motor control, with the hip gradually becoming the reference joint for balance during gait. Hip stability develops progressively, enabling well-regulated, anticipatory walking. The child then abandons step-by-step advancement and fall recovery, where double support alone ensured precarious stabilization (Bril et al. 1990). Once this dynamic mastery is acquired, segmental fluidity between body parts emerges (Bril et al. 1993), typically around 46 weeks after autonomous walking begins. At this stage, the child can regulate cyclical imbalances in gait through efficient sensory integration, fluid transitions between sensory modalities and shifts between visual and proprioceptive engagement. In adults, small head accelerations observed during walking are directly linked to hip torque adjustments, occurring without sensory system intervention (Winter et al. 1993).

In 6-year-old children with advanced gait expertise, tasks requiring greater balance (e.g. walking on a beam) necessitate head-trunk fixation as a block (Assaiante et al. 1993). After age 9, the head is anchored in space with relative independence from trunk movements, and visual control becomes the primary factor in head stabilization. Head movements then perfectly maintain trunk balance (Assaiante et al. 1995). In children with cerebral palsy, head stabilization differs from that of able-bodied peers (Wallard et al. 2012); robot-assisted gait training can reorganize this pattern (Wallard 2014). Ocular proprioception linked to oculomotor activity and cervical proprioception linked to vestibular input also provide information on head stabilization, each offering egocentric references for orientation in extracorporeal space (Roll et al. 1987).

In both reported cases, compensation through the best-preserved sensory modality can be assumed.

In Jeremy, head stabilization is cervid-like, with a sagittal trunk axis that does not respect physiological curvatures. Under these conditions, the axis of the horizontal semicircular canals is misaligned, possibly leading to vestibular stimulations different from those of a child who maintains the orbitomeatal axis aligned with distant gaze. To our knowledge, no studies have examined this specific issue. Clinically, Jeremy's visual and proprioceptive capacities are preserved. He demonstrates strong oculomotor and visuospatial abilities, mastery of body organization and analytical control of his muscles, including biarticular ones. By age 8 to 10, in a sitting position, he can regain perfect vertical trunk alignment from pelvis to vertex in both sagittal and frontal planes, provided the effort is not sustained.

In Lisa, visual function is overstimulated and captures all information necessary for deliberate sequencing of motor activities. Even a fleeting glance provides effective guidance for optimal step landing, enabling pelvic balance control (Assaiante et al. 1989).

Other factors also contribute to posture. Although less well-known, they may become predominant in certain circumstances. It is possible that these inputs converge at the neuromuscular spindle, informing it directly (Johansson et al. 1991) or sending copies of information to the central nervous system (Leroy-Malherbe 2018). Tactile factors play a role in triggering posture through changes in ground support (Couillandre et al. 2003; Roll 2003) or plantar tactile stimulation (Do et al. 1990; Kanavounoudias et al. 2001). Tactile receptors are concentrated around the big toe and the outer edge of the sole (Kennedy et al. 2002). Sensitive to skin stretch, they integrate spatiotemporal and frictional data from the sole, playing a fundamental role in reflexes attributed to the central gait generator that drives step advancement (Dietz 1992). As seen in automatic walking in newborns, which already contains the biomechanical characteristics of future gait (Bril et al. 1994), step alternation and sequencing are triggered and maintained by plantar tactile stimulation. The landing of the swinging step, crucial for lateral balance, also engages plantar cutaneous afferents. These stimulations initiate postural straightening and reflex alternation of lower-limb swaying. A field of cutaneous receptors also exists along the lateral malleolus, activated during sudden varus destabilization of the ankle. In such cases, skin stretch at the malleolus elicits faster responses than neuromuscular spindle stretch, making it more effective in restoring balance (Perry et al. 2000).

In subjects with cerebral palsy, podiatric care and footwear conditions are therefore critical. Rehabilitation using compression suits or strapping has shown that sustained tactile stimulation can improve straightening of body segments (Sancelme 2021; Genin Journet 2023).

Among other sensory modalities, even auditory detection can incidentally contribute to posture, which explains the interest in "sound space" during rehabilitation.

7.3.2.3. *Sensory integration*

Sensory integration refers to the system of modulation, cancellation or summation of signals from multiple sensory receptors stimulated for the purpose of motricity (Haggard et al. 2013). This weighting of information intake allows complex movements to remain harmonious, ensuring a lived stability of the body.

Take, for example, a gymnastics exercise such as the somersault. During rotation, the gymnast cancels visual input in favor of vestibular information; at the moment of ground reception, visual and proprioceptive inputs dominate over vestibular ones. Thus, a successful somersault requires continuous shifts in sensory prioritization. The ability to transition from one sensory selection to another is not systematically understood in subjects with motor disabilities, and to our knowledge, no study has focused on this competence in such populations.

Sensory integration involves several modalities, each with its own filters, propagation speed and detection latency (Leroy-Malherbe 2008). For the same movement, the gain of each sensory input can vary depending on ground and environmental conditions, the type of motor program triggered, the subject's expertise and age (Woollacott et al. 1987). Each sensory process thus has a specific detection role while also contributing transversally. For instance, tactile receptors provide information on verticality from the bottom upward, whereas vestibular receptors provide information from the top downward. Vestibular input, dependent on head movements, detects sudden changes requiring rapid regulation, while proprioception, which encompasses the whole body, conveys information on slower movements or static balance. The vestibular system produces long-lasting responses, whereas proprioception generates brief reactions that must be repeatedly triggered to remain active.

These distinctions in sensory detection are significant for spatial exploration and perhaps even for psychological processes. In adolescents, transient neglect of proprioceptive input in favor of visual information has been observed (Viel et al.

2009), raising the question of whether such prioritization serves as psychological protection during periods of turbulent bodily change. All sensory inputs converge in the cortex, where they are integrated within associative areas. Vision alone mobilizes 30% of cortical resources, compared to 10% for touch and 3% for hearing. This subtle dialogue among sensory modalities enables harmonious, well-coordinated motor activity and reduces detection ambiguities (Carver et al. 2006). Although the ability to select sensory inputs evolves throughout development until late adolescence (Assaiante 2016), it has not been studied in children with motor disabilities.

This raises the question of how the two children presented here manage sensory information.

In Jeremy's case, given his slow but steady progress towards upright posture despite lacking automatic reactions to gravity, we might consider that he overstimulates the necessary sensory modalities to develop postural abilities wisely. He likely alternates or integrates these modalities optimally according to his needs, ensuring no loss of information detection and potentiating the effects of accumulated inputs to compensate for deficient postural reflexes. It is possible that additional sensory resources, typically recruited only at specific life stages, are also mobilized in him. Notably, the therapeutic measures applied were minimal ("untiltable" footwear with a large sole, a single toxin injection and rehabilitation focused on neck and posture). Yet his motor gains exceeded expectations: walking unaided outdoors and regulating stride according to environmental conditions. This fragile construction of upright motor skills explains the surgeon's caution regarding surgical intervention.

For Lisa, organizational strategies in the absence of any perceived reference of body position prioritized vision for movement detection and control, recalibrating other modalities that are normally secondary in posture. Hearing, which can activate a "sound space" and provide a body envelope in cases of visual deficit, may contribute to her spatial detection in the absence of proprioception. However, the type of "subclinical" detection she develops is more likely linked to two cognitive factors involved in posture – factors central to training in human movement with artistic aims, such as dance.

7.4. Integrative factors affecting posture

Two additional hypotheses concerning integrative factors have caught our attention: one relates to the emotional sphere, and the other to the psychosocial sphere. The first idea was originally advanced by dance practitioners such as Hubert Godard (Godard 1998) – figures engaged in extraordinary bodily practice, where automatism is acquired only through rigorous and highly tactical work on bodily input and output. The second idea was defended by Marcel Mauss (Mauss 2021), an observer of the body during interactive social exchanges. His point of view describes the familiar situation of the body as a cultural and anthropologic coherent practice of gift and counter-gift. These authors highlight the symbolic functions of verticality, expressed through the internalized body image or through the social body.

It is therefore not inappropriate to compare our two children to extraordinary individuals – those for whom walking has become an art and who summon the same ambition of bodily discipline in the face of profound deficits.

7.4.1. *The tonic-emotional body*

Hubert Godard was particularly interested in interpreting movement in dancers to theorize a capacity for muscle control that integrates the imbalances produced by movement throughout the body. For example, a dancer who moves their arm will, in the preceding seconds, experience an automatic contraction of the calves and thus of the lower body, precisely anticipating the planned change in the upper body. This observation is confirmed by the existence of anticipatory postural adjustments, detectable through electromyography of postural muscles or their kinetic effects. The contraction that precedes movement contains all the information about the forthcoming action: analysis of the electromyography of the pectoral girdle a few seconds before lifting dumbbells can predict whether a weightlifter's snatch will succeed; similarly, the recoil of the center of pressure just before a sprint can indicate the sprinter's top speed (Bouisset 1995; Noé 2016). The visible position change is only the surface; the anticipatory tonic function, though invisible, reveals everything about the upcoming movement. In this anticipatory function, Godard emphasizes the importance of mood and affect in the dancer, linking inner stability with outer stability. Kinematic studies support this hypothesis (Berthoz 2013).

For Godard, anticipatory postural function inscribes in the body successive tension shifts – a memory of the subject's history embedded in spatially and temporally organized bodily relationships. It integrates the body in motion while unfolding the subject's motor history. In dance, this means linking the tonic value exerted by muscles that maintain limb segments and support the body against gravity (posture preparing movement) with the gesture itself, which unfolds in time (temporality) and space (form) against this tonic background. The organization of limb segments in time and space creates a kinetic melody, reflecting the personality of the evolving subject. Psychomotor therapists call this the tonic-emotional dialogue.

Though these ideas have long remained theoretical, a biological discovery may support them: titin, a protein abundant in muscle, links actin-myosin complexes. By opposing sarcomere stretch, titin retains a form of muscle tension memory and, through biochemical reactions, adjusts elasticity according to training (Ferreiro 2017). This memory of resisted tension is precisely what is mobilized in practices such as yoga.

The link between the body as felt and the body as acted upon across the subject's life establishes what psychomotor therapists call body image. Movement is no longer merely mechanical but symbolic, carrying meaning in relation to the world: body in motion – identified body; anticipated movement – historicized body. This tonic-emotional dialogue is clinically observable even in infants carried by adults and originates in intra-uterine life: the baby is carried by its mother, then later, separated, becomes carried by Earth's gravity. Godard highlights the continuity between these phases – maternal carrying and autonomous maintenance. Catherine Dolto, co-author of this book, emphasizes the parallel between "being carried" and "carrying oneself." The way a baby adjusts in the arms of loved ones expresses reciprocal tenderness (affective tuning) perceived since life in utero, accompanying the ability to prepare movement through adapted tone (postural anticipation). The form of movement, beginning in anticipation, becomes a trace of affective connection.

Godard observed in dancers how the same movement scheme, such as arm extension, can differ profoundly: in some, anticipatory contraction of the serratus anterior marks a pushing gesture; in others, contraction of the pectoralis major marks a catching gesture. Depending on which muscle initiates the gesture, execution acquires diametrically opposed qualities and effects, even though the trajectory is identical. To launch a shot is to push; to launch a javelin is to throw. Performing a jeté in dance requires bodily understanding of the nuances between pushing, launching and throwing, activating muscle patterns that best convey the intended expression.

These elements of bodily language underpin stylistic variations and the dancer's connection with the audience. The characteristics of anticipatory gesture expression are deeply personal experiences of Earth's gravity. Gravity, according to Godard, is the founding reference for our relationships, shaping our way of being with our bodies, the environment and others. Stress or trauma can disrupt this body image, breaking the continuity between "being carried by the mother" and "carrying oneself through gravity." The body as felt no longer aligns with the body as acted upon; the memorized infant body disconnects from the ability to maintain oneself in action. Gravity becomes oppressive rather than a vector of movement, forcing the subject to struggle against it instead of moving gracefully through it.

Can we suggest that the affective tuning of our two patients – sustained by family and personal histories that preserve inner stability despite challenges – provides the foundation for their motor development? Do they draw primarily on this early tonic-emotional dialogue, ensuring continuity in the harmonious use of gravity? Does their particularly solid body image allow fluidity between early lived experience of gravity and its natural use, even beyond severe neurological impairment? Could this compensatory functioning be possible only because the neurological deficit of an essential function remains isolated?

In Jeremy's case, cognitive spatialization, ocular mobility and intellectual functioning are highly competent.

In Lisa's case, intellectual vitality, desire to move and awareness of her immediate material environment are strongly developed.

The enigma posed by long-term monitoring of these two children remains unresolved. Conventional analysis of their neurological impairment does not suffice to explain their progress. We must add one final hypothesis: that of the social body as a means of negotiating gravity.

7.4.2. *The social body*

It was Marcel Mauss who, in his inaugural conference of 1936, invited us to consider the gesture of cultures before that of bodies or functionalities. According to him, motor skills as fundamental as walking or swimming hold a status specific to each people and each era. This ethnography of gesture highlights the shared character of bodily movement within a broad societal group. For Mauss, it is a matter of engraining a bodily technique – a technique the child acquires by imitating what it perceives as successful in society, with trusted figures, such that the act of imitation is imposed from the outside, even for biological acts concerning the body. Education thus intervenes in the use of the human body: the way people squat on the roadside in China, the swaying of the hips in Maori walking, or the African practice of carrying a baby on the back – all stem from communal learning. This use of the body as a cultural tool refers to the body schema 1 and can be likened to the terminology of praxis (Leroy-Malherbe 2017b, 2017c, 2017d).

Might standing upright be conceived as a gesture so valued and encouraged that symbolic construction comes to support – or even replace – neurological wiring? Could the concept of praxis, usually applied to fine motor skills, be extended to the gross motor skill of walking? In situations of motor disability, might the body as a whole be mobilized as thoughtfully and as finely tuned as the hands of an able-bodied subject?

If this hypothesis of "gross motor skills" related to praxis under certain conditions remains valid, the role of imitation in motor learning can be invoked in the two cases we report. School, as a social group, is already recognized as a place for modeling gesturality in children (learning to write, for example) or for shaping body image in adolescents (through fashion phenomena or tattoos). It is also a place for modeling gross motor skills in children with intellectual or motor deficits (Leroy-Malherbe 2010). For these children, gross motor skills are learned step by step, with a model, much as fine motor skills are learned by able-bodied children.

We therefore hypothesize that daily school attendance and the observation of peers provided Jeremy and Lisa – whose clinical situations we have described – with strong models of how their bodies should be held at their age. Their eagerness to conform to these models may have transformed them into genuine tutors for their naturally deficient posture.

7.5. Conclusion

"The miracle is not to walk on water. The miracle is to walk on the green earth, dwelling deeply in the present moment, and feeling truly alive," says ancient Asian wisdom. In a very real and non-miraculous way, our two patients defied medical examination conducted according to the orthodoxy of current teaching.

"Walking is an experience of indefiniteness and infinity. With each step forward, one receives, wide open and deep within oneself, the gifts of the universe," continues the text displayed along the winding slope at the Musée du Quai Branly in Paris, as one enters the collections of different cultures. What better way to express the link between walking and Earth's gravity, permitted by our universe? And what better way to highlight the breadth of resources that sustain upright posture beyond the known decisive factors – when neither adapted control of the stretch reflex nor proprioceptive information provides support?

Where does the subject draw this essential ability from? From self-determination? From the strength of the social model? Or from symbolic dimensions of verticality? It is likely that all these immeasurable resources contribute to reaching the fundamentals of human experience: the force of gravity and the reality of verticality.

Indeed, for every living being – plant, animal or human – the point of penetration of the male gamete into the female gamete directs the development of the future individual, determining both orientation and trajectory. In plants, this point corresponds to the roots; in animals, to the head (Halle 1999). From conception, the living being carries within itself what is needed to confront Earth's gravity. This also implies the disappointment of those who fail to meet these challenges and the caution that our words as physicians must respect in the face of the expectations of the young people we accompany (Truscelli 2008).

7.6. Acknowledgments

On the eve of her retirement, after 36 years devoted to physical medicine and the rehabilitation of brain-injured children, Véronique Leroy-Malherbe wishes to express her gratitude, through this literature review, to all those who have contributed depth and coherence to her professional journey – particularly Danièle Truscelli, Yvon Brenière, Pierre Landrieu, Philippe Toulet, Georges Thuilleux and the SESSAD teams of APF France Handicap.

7.7. References

Amblard, B., Assaiante, C., Fabre, J.C., Mouchnino, L., Massion, J. (1997). Voluntary head stabilization in space during oscillatory trunk movements in the frontal plane performed in weightlessness. *Exp. Brain Res.*, 114, 214–225.

Assaiante, C. (2016). Développement perceptivo-moteur de l'enfant et de l'adolescent et contrôle postural. In *Posture et équilibration humaines*, Paillard, T. (ed). De Boeck, Louvain, 164–165.

Assaiante, C. and Amblard, B. (1993). Ontogenesis of head stabilization in space during locomotion in children: Influence of visual cues. *Exp. Brain Res.*, 93, 499–515.

Assaiante, C. and Amblard, B. (1995). An ontogenic model for the sensori-motor organization of balance control in humans. *Hum. Mov. Sci.*, 14, 13–43.

Assaiante, C., Marchand, A.R., Amblard, B. (1989). Discrete visual samples may control locomotor equilibrium and foot positioning in man. *J. Mot Behav.*, 21(1), 72–91.

Assaiante, C., Barlaam, F., Cignetti, F., Vaugoyeau, M. (2014). Building of body schema during childhood and adolescence: A neurosensory approach. *Neurophysiol. Clin.*, 44, 3–12.

Barale, M.E. and Macqueron, A. (2016). Apport des panneaux Vichy dans l'acquisition de la station assise depuis son utilisation au CAMSP en 2011. *Mot. Cer.*, 37(3), 102–107.

Berthoz, A. (2013). *Le sens du mouvement*. Odile Jacob, Paris.

Bouisset, S. (1995). Stabilisation de la posture. In *Muscles, posture et mouvement : bases et applications de la méthode électromyographique*, Bouisset, S. and Maton, B. (eds). Hermann, Paris, 497–498.

Bouvier, J. (2014). Neurons in motion: Spinal networks controlling left-right alternation of walking movements deciphered by genetics. *Med. Sci.*, 20, 121–124.

Brenière, Y. (1998). Effets posturo-cinétiques de la fréquence des pas sur l'équilibre dans le plan frontal au cours de la marche. In *STAPS. Sport, posture et équilibration*, Crémieux, J. and Rougier, P. (eds). De Boeck, Louvain, 31–40.

Brenière, Y. and Do, M.C. (1986). When and how does steady-state gait movement induced from upright posture begin? *J. Biomech.*, 19(12), 1035–1040.

Brenière, Y. and Do, M.C. (1991). Control of gait initiation. *J. Mot. Behav.*, 23(4), 235–240.

Brenière, Y., Do, M.C., Bouisset, S. (1987). A biomechanical study of the gait initiation process. *J. Biophys. Méd. Nucl.*, 5(4), 197–205.

Brenière, Y., Bril, B., Fontaine, R. (1989). Analysis of the transition from upright stance to steady state locomotion in children with under 200 days of autonomous walking. *J. Mot. Behav.*, 21(1), 20–37.

Bril, B. and Brenière, Y. (1990). How does the child modulate progression velocity during the first 18 months of independant walking? In *Disorders in Posture and Gait, X[th] Int. Symp. Soc. for Postural and Gait Research*, Munich.

Bril, B. and Brenière, Y. (1992). Postural requirements and progression velocity in young walkers. *J. Mot. Behav.*, 24(1), 105–116.

Bril, B. and Brenière, Y. (1993). Posture and independant locomotion in early childhood. In *The Development of Coordination in Infancy*, Salvelsbergh, G.J.P. (ed.). Elsevier, Amsterdam, 337–358.

Bril, B. and Ledebt, A. (1994). Les débuts de la marche autonome chez l'enfant. *J. Pédiatrie Pueric.*, 7(1), 21–33.

Bruijn, S.M. and Van Dieen, J.H. (2018). Control of human gait stability through foot placement. *J.R. Soc. Interface*, 15, 143.

Bruyneel, A.V. (2022). Évaluation de la proprioception. *E.M.C.*, Elsevier Masson, Paris, 1–13.

Bullinger, A., de Santa Ana, I., Grival, P., Millan, R., Scheidegger, P., Schmud Pons, N., Tschopp, C. (1996). Le bilan sensori-moteur de l'enfant : éléments théoriques et cliniques. In *Enfances*. PUF, Paris, 41–50.

Carver, S., Kiemel, T., Jeka, J.J. (2006). Modeling the dynamics of sensory reweighting. *Biological Cybernetics*, 95(2), 123–134.

Collins, S., Ruina, A., Tedrate, R., Wiss, M. (2005). Efficient bipedal robots based on passice-dynamic walking. *Science*, 307, 1082–1085.

Coppens, Y. (2000). *Le genou de Lucy : l'histoire de l'homme et l'histoire de son histoire*. Odile Jacob, Paris.

Couillandre, A. and Brenière, Y. (2003). How does the heel-off posture modify gait initiation parameter programming? *J. Mot. Behav.*, 35(3), 221–227.

Dietz, V. (1992). Human neuronal control of automatic functional movements: Interaction between central programs and afferent input. *Physiol. Rev.*, 72, 33–69.

Do, M.C., Bussel, B., Brenière, Y. (1990). Influence of plantar cutaneous afferents on early compensatory reactions to forward fall. *Exp. Brain Res.*, 79, 319–324.

Duval-Beaupère, G. and Robain, G. (1989). Les rapports anatomiques du point d'application de la masse du segment corporel supporté par chaque étage vertébral. *Ann. Kinesithér.*, 16(1/2), 1–7.

Elftman, H. (1939). Forces and energy changes in the leg during walking. *Am. J. Physiol.*, 125, 339–356.

Eng, J.J. and Winter, D.A. (1995). Kinetic analysis of the lower limb during walking: What information can be gained from a three-dimensional model. *J. Biomech.*, 28(6), 753–758.

Ferreiro, A. and Urtizberea, J.A. (2017). Pathologies musculaires liées à la titine : un domaine émergent. *Med. Sci.*, 33, 16–21.

Genin Journet, M. (2023). Un compromis dans l'appareillage pour soutenir les activités de la vie quotidienne : l'orthèse Dynamique Flexa. *Mot. Cer.*, 44(2), 44–54.

Godard, H. (1998). Le geste et sa perception. In *La danse au XXe siècle*, Michel, M. and Ginot, I. (eds). Bordas, Paris, 224–229.

Haggard, P., Iannetti, G.D., Longo, M.R. (2013). Spatial sensory organization and body representation in pain perception. *Curr. Biol.*, 18, 23.

Halle, F. (ed.) (1999). Voyage au pays de la forme. In *Éloge de la plante : pour une nouvelle biologie*. Points, Paris, 81–83.

Johansson, H., Sjolader, P., Sojka, P. (1991). Receptors in the knee joint ligaments and their role in the biomechanics of the joint. *Clin. Orthoped.*, 268, 161–178.

Kanavounoudias, A., Roll, R., Roll, J.P. (2001). Foot sole and ankle muscles inputs contribute jointly to human erect posture regulation. *J. Physiol.*, 538(3), 869–878.

Kennedy, P.M. and Inglis, J.T. (2002). Distribution and behaviour of glabrous cutaneous receptors in the human foot sole. *J. Physiol.*, 538(3), 995–1002.

Kiehn, O. (2011). Development and functional organisation of spinal locomotion circuits. *Curr. Opin. Neurobiol.*, 10, 507–518.

Kuo, A.D. (1999). Stabilization of lateral motion in passive dynamic walking. *Int. J. Rob. Res.*, 18, 917–930.

Le Métayer, M. (1999). *Rééducation cérébro-motrice du jeune enfant : éducation thérapeutique*. Elsevier Masson, Paris.

Ledebt, A., Bril, B., Wiener-Vacher, S. (1995). Head coordination in the sagittal plane in toddlers during walking: Preliminary results. *Neuroreport*, 6, 1737–1740.

Leroy-Malherbe, V. (2008). Cognition et motricité chez l'enfant infirme moteur cérébral : quelles perspectives pour la recherche clinique ? In *Les infirmités motrices cérébrales : réflexions et perspectives sur la prise en charge*, Truscelli, D. (ed.). Elsevier Masson, Paris, 295–325.

Leroy-Malherbe, V. (2010). Orientation et organisation de la scolarité du jeune IMC. *Mot. Cer.*, 31(1), 98–115.

Leroy-Malherbe, V. (2017a). Examen médical neuro-moteur. In *Comprendre la paralysie cérébrale et les troubles associés. Évaluations et traitements*, Truscelli, D. (ed.). Elsevier Masson, Paris, 89–114.

Leroy-Malherbe, V. (2017b). L'imagerie mentale : faits culturels dans l'écriture, la peinture et la poésie. *Mot. Cer.*, 38(3), 87–97.

Leroy-Malherbe, V. (2017c). Arguments pour dynamiser l'émergence de l'imagerie mentale auprès d'enfants porteurs de pathologies neurologiques centrales. *Mot. Cer.*, 38(2), 41–49.

Leroy-Malherbe, V. (2017d). L'enrichissement de l'imagerie mentale comme invitation au développement de la cognition. *Mot. Cer.*, 38(1), 1–12.

Leroy-Malherbe, V. (2018). Posture, équilibre et mouvement : rôle de la proprioception dans la verticalité. *Mot. Cer.*, 39(1), 4–12.

Leroy-Malherbe, V., Brenière, Y., Bril, B. (1992). Comment l'enfant atteint d'hémiplégie cérébrale infantile contrôle-t-il sa marche ? *Ann. Réadapt. Med. Phys.*, 35, 433–438.

MacGeer, T. (1990). Passive dynamic walking. *Int. J. Rob. Res.*, 9, 62–82.

MacKinnon, C.D. and Winter, D.A. (1993). Control of whole body balance in the frontal plane during human walking. *J. Biomech.*, 26(6), 633–644.

Mauss, M. (2021). *Les techniques du corps*. Payot/Actes Sud, Paris.

Noé, F. (2016). Les ajustements posturaux anticipés. In *Postures et équilibration humaines*, Paillard, T. (ed.). De Boeck, Louvain, 113–118.

Perry, S.D., MacIlroy, W.E., Maki, B.E. (2000). The role of plantar cutaneous mechanoreceptors in the control of compensatory stepping reactions evoked by unpredictable multidirectional perturbation. *Brain Res.*, 877, 401–406.

Pohl, M.B., Kendall, K.P., Petel, C., Wiley, J.P., Emery, C., Ferber, P. (2015). Experimentally reduced hip-abductor muscle strength and frontal-plane biomechanics during walking. *J. Athl. train.*, 50, 385–391.

Redon-Zouiteni, C., Roll, J.-P., Lacert, P. (1994). Reprogrammation posturale d'origine proprioceptive chez l'enfant paralysé cérébral. *Mot. Cer.*, 15–20.

Roll, J.-P. (2003). *Pied, équilibre et traitements posturaux*. Masson, Paris.

Roll, J.-P. and Roll, R. (1987). Rôle de la proprioception musculaire somatique et extra-oculaire dans la régulation posturale et le codage du mouvement. *Mot. Cer.*, 8, 49–57.

Roll, J.-P., Vedel, J.-P., Gilhodes, J.-C., Ribot, E. (1986). Proprioceptive sensory coding of motor activities in man. In *Sensorimotor Plasticity*, Ron, S., Schmid, R., Jeannerod, M. (eds). INSERM, Paris, 283–304.

Sala, D.A. and Grant, A.D. (1995). Prognosis for ambulation in cerebral palsy. *Dev. Med. Child Neurol.*, 37(11), 1020–1026.

Sancelme, Z. (2021). Rôle proprioceptif du body compressif dans la rééducation d'un enfant paralysé cérébral. *Mot. Cer.*, 42(1), 2–6.

Tardieu, C. (2018). Origine et évolution de la bipédie. *Mot. Cer.*, 39, 55–65.

Truscelli, D. (2008). Problèmes médicaux du diagnostic et du pronostic locomoteur. In *Les infirmités motrices cérébrales : réflexions et perspectives sur la prise en charge*, Truscelli, D. (ed.). Elsevier Masson, Paris, 59–69.

Truscelli, D. (2017). Introduction. Pronostic locomoteur. In *Comprendre la paralysie cérébrale et les troubles associés. Évaluations et traitements*, Truscelli, D. (ed.). Elsevier Masson, Paris, 1–4.

Vasseur, R. (2000). Importance des aspects biomécaniques et des points d'appui posturaux dans la genèse de l'axe corporel. In *Enfance : le bébé, le geste et la trace*, Bullinger, A. (ed.). PUF, Paris, 221–233.

Viel, S., Vaugoyeau, M., Assaiante, C. (2009). Adolescence: A transient period of proprioceptive neglect in sensory integration of postural control. *Motor Control*, 13, 25–42.

Wallard, L. (2014) Étude de la marche avant et après une rééducation robotisée chez des enfants présentant des troubles locomoteurs. Thesis, Université Paris Descartes, Paris.

Wallard, L., Bril, B., Dietrich, G., Kerlirzin, Y., Bredin, J. (2012). The role of head stabilization in locomotion in children with cerebral palsy. *Ann. Phys. Rehab. Med.*, 55, 590–600.

Wallard, L., Dietrich, G., Kerlirzin, Y., Bredin, J. (2014). Balance control in gait children with CP. *Gait Posture*, 40(1), 43–47.

Wallard, L., Dietrich, G., Kerlirzin, Y., Bredin, J. (2016). Apprentissage moteur et contrôle postural chez l'enfant. In *Postures et équilibration humaines*, Paillard, T. (ed.). De Boeck, Louvain, 183–189.

Wang, Y. and Srinivasan, M. (2014). Stepping in the direction of the fall: The next foot placement can be predicted from upper body state in steady-state walking. *Biol. Lett.*, 10.

Winter, D.A. (ed.) (1990). Mechanical work, energy and power. In *Biomechanics and Motor Control of Human Movements*, 2nd edition. Wiley, Waterloo, 103–139.

Winter, D.A. (1995). Human balance and posture control during standing and walking. *Gait Posture*, 3, 193–214.

Winter, D.A., MacKinnon, C.D., Ruder, G.K., Wieman, C. (1993). An integrated EMG/ biomechanical model of upper body balance and posture during human gait. *Prog. Brain Res.*, 97, 359–367.

Witcher, R. (2019). Editorial. *Antiquity*, 93(368), 287–296.

Woollacott, M., Debu, B., Mowatt, M. (1987). Neuromuscular control of posture in the infant and the child: Is vision dominant? *J. Mot. Behav.*, 19, 167–186.

Yang, F. and Pai, Y.C. (2014). Can sacral marker approximate center of mass during gait and slip-fall recovery among community-dwelling older adults. *J. Biomech.*, 47, 3807–3812.

8

Contribution to the Study of the Postural Evolution of Patients in a Dental Office

8.1. Introduction

Paleoanthropology has led us to discover that dental occlusion became a parameter of verticality and therefore a postural clue. Dambricourt Malassé has demonstrated and modeled since 1987 the changes in morphological relationships between the face and the skull base over the course of its straightening since the first primates, a dynamic process that she calls "craniofacial contraction", and generated over the course of embryogenesis of the neural tube (Dambricourt Malassé 2022a, 2022b). These relationships came to be corroborated by studies in dento-maxillofacial orthopedics (DFO) dedicated to the three skeletal and dental classes, focusing on spheno-occipital synchondrosis (Deshayes 1986, 1991) and their application in paleoanthropology (Hadjouis 2021). This holistic intuition has been with us since 1980, when we studied osteopathy, then posturology, and then in 1992 defined the concept of orthoposturodonty, which aimed to treat postural deficits using occlusal treatment (Clauzade and Darraillans 1989, 1992; Clauzade and Marty 1998). We thought we had worked for the good of our patients, because the results were spectacular, but we clashed with the academic dogmas and ordinals of our profession. We believe today, from statistics made with our dental, osteopathic and podiatrist participants, that 40% of them present with postural deficits of occlusal origin, most often unknowingly.

Chapter written by Michel CLAUZADE and Numa CLAUZADE.

8.2. The trigeminal nerve: neuro-matrix axis of the face

However, while this osteocranial architectonic observation can be noted in paleoanthropology and paleontology, which work with fossils, it does not satisfy us as practitioners working with living organisms.

Le Double suggested back in 1903 that brain development certainly played a part in the morphology of the skull's bones, and he foresaw the pre-eminent role of the encephalon and its sensory (the placodes) or sensitive (the cranial nerves) expansions, in the determinism of the emergence of bony parts (Le Double 1903; Clauzade and Marty 1998, 2006). Ossification observes a topographical and chronological schematization. The sites of emergence, called ossification centers, follow a path corresponding to the nerve trajectory, while the growth of the bony piece is governed by dynamic phenomena that are predominantly muscular but with, at least indirect, nerve obedience. Bone emergence seems to correspond to a nervous, sensitive or sensory presence introducing the concept of the neuro-sensory axis, advanced by Laude, who presented it at the mandible level; thus, the trigeminal constitutes the neuro-matrix axis of the face (Clauzade and Marty 1998, 2006). The trigeminal, a nerve of the first branchial arch, will offer neural models with orthogonal orientation to mesenchymal, then cartilaginous, ossification. The neural matrix constitutes the architectural reference for mandibular orientation relative to the skull base (Stricker and Raphael 1993).

Damasio (1999) describes the trigeminal nerve as a global nerve because unlike other cranial nerves, it can convey the three types of somatosensory information that the brain collects:

– visceral-internal or interoceptive information;

– musculoskeletal or exteroceptive information;

– fine-discriminative or proprioceptive information.

To be added to this is the motor information provided by its V3 branch at the musculoskeletal and sensory level (*chorda tympani*).

The essential point to be emphasized is that the trigeminal nerve (Figure 8.1), through its neurosensory and motor properties, participates in the brain's mediating role between the internal and external worlds (Clauzade and Marty 2006; Clauzade and Clauzade 2016).

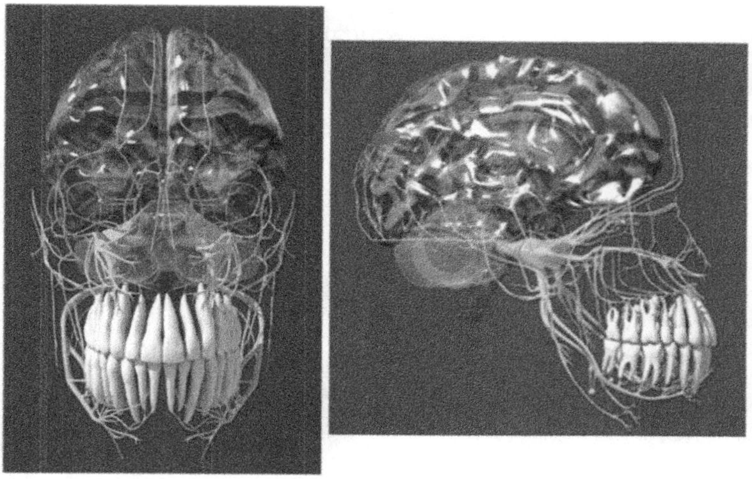

Figure 8.1. *The trigeminal nerve – front and profile views. For a color version of this figure see www.iste.co.uk/dambricourtmalasse/futureofsapiens.zip*

"When the wise man points at the moon, the idiot looks at the finger": the moon represents the trigeminal and the finger represents dental occlusion. Posture is a dynamic act, which already requires a lot of energy to fight the force of gravity. It is the role of the postural tonic system that allows straightening and that is of noradrenergic origin (trigeminal-adrenal) (Clauzade and Clauzade 2016; Bricot 2020). We suggested in 1989 that dental occlusion was an "open suture", a place of escape from cranial constraints and therefore a print, a fossilization, of the physical and emotional functioning of the individual (Clauzade and Daraillans 1989). For us, the trigeminal is the guiding thread in understanding between skull and posture. The error is to focus on the V1, V2 and V3 branches after the exit from the Gasser ganglion and to forget the intra-cranial part. The trigeminal emerges from the lateral pontine part of the brain stem, but it originates from a strip of three nuclei that are positioned in the dorsal part of the reticular formation and spread from the diencephalon to the elongated marrow, i.e. at cervical vertebra C2. These trigeminal nuclei (Figure 8.2) will interconnect with different nuclei of the reticular formation, which play a fundamental role in our behavioral and emotional relationship life: *locus coeruleus*, noradrenergic center; *tractus solitarius*, cardio-respiratory center; para-brachial (PB) nucleus and periaqueductal gray matter (PGM), management of pain and emotions; and vestibular, nuclei postural role (Clauzade and Clauzade 2006).

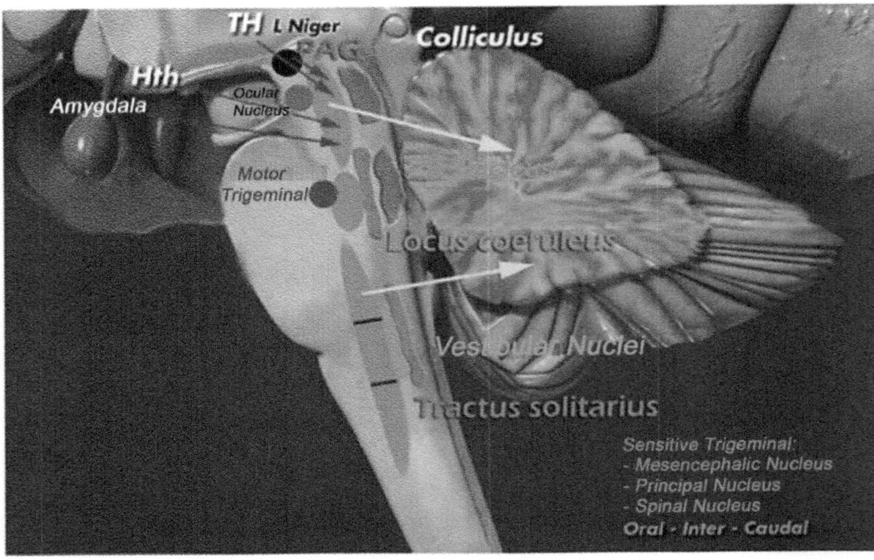

Figure 8.2. *Trigeminal nuclei. For a color version of this figure see www.iste.co.uk/dambricourtmalasse/futureofsapiens.zip*

These three sensitive trigeminal nuclei have different functions, to which should be added the motor nucleus, very close to the main nucleus, which can constitute a very efficient sensory-motor reflex arc (Stricker and Raphael 1993):

– The mesencephalic nucleus: located in the diencephalon, it is a veritable integrating center that receives information from the hypothalamus, thalamus, amygdala, *locus niger* and likely from the *area postrema* of the 4th ventricle. Its projections will go onto the cerebellum and the colliculi. It is responsible for the fine and discriminative proprioception of the teeth and the periodontium (Strazielle 1996). It controls the masticatory muscles and the trigeminal-masseter reflex (Stricker and Raphael 1993). Its relationship with the ocular and vestibular nuclei explains its importance in posturology, in the vestibulo-ocular-cephalogyric reflex. On the emotional level, it will govern aggression and attack-and-defense behaviors, and painful sensations, crying and tears associated with the PAG nucleus.

– The principal nucleus: plays the role of the dorsal column of the somatosensory system and conveys the information of touch, pressure, vibration and the conscious sense of joint positions.

– The spinal nucleus: regulates the joint-boundary and large-amplitude movements of temporomandibular joints (TMJs) and participates in periodontal proprioception. It contributes above all to the perception of pain, its mediation and its quality, associated with the PB nucleus. It contributes to the phenomena of nausea and unpleasant taste and food motivation. It has many connections with the cerebellum (Strazielle 1996; Ito 2001).

This interceptive information provided by the trigeminal offers us a clearer understanding of the link with patients' posture and the acquisition of verticality. Other later elements will reinforce this relationship.

However, we have already mentioned the increasing number of patients we encounter for postural deficits, and lumbar pain is becoming a health issue. We can therefore ask ourselves the legitimate question of whether modern man is always adapted to this gravity; otherwise, what would be the causes of this mismatch?

8.2.1. *Statistical study*

In 2016, we carried out a statistical study at a dental office using the clinical records of 853 patients treated over six years, in the context of craniomandibular pathologies or temporomandibular dysfunctions, including postural deficits from which these individuals suffered, knowingly or not (Clauzade and Clauzade 2006). The analysis of the results revealed undeniable factual elements to us.

8.2.1.1. *Gender distribution*

The prevalence of female patients is indisputable; they represent 70% (69.26) of cases (Figure 8.3). How should this gap be analyzed? Women already attend dental offices more regularly for aesthetic or cosmetic reasons, but in the present case, our consultations focus on pain, which does not appear to be an argument that is to be retained.

They are much more likely to present with temporomandibular dysfunctions, as well as migraines, cephalalgia or fibromyalgia. Estrogens have been widely evoked in the emergence of these disorders. Some suggest autoimmune rheumatoid diseases that affect multiple joints (Bracco et al. 2010) or systemic rheumatoid diseases, often autoimmune, such as lupus or scleroderma (Clauzade and Clauzade 2006; Lin et al. 2007; Carpentier and Felizardo 2012), which are more common in women: five patients presented with autoimmune diseases, but this number is small in relation to our sampling. Carpentier suggested that the structure of type-II present in collagen in TMJs is very sensitive to decreased estrogen levels, especially during menopause (Carpentier and Felizardo 2012) (Figure 8.4).

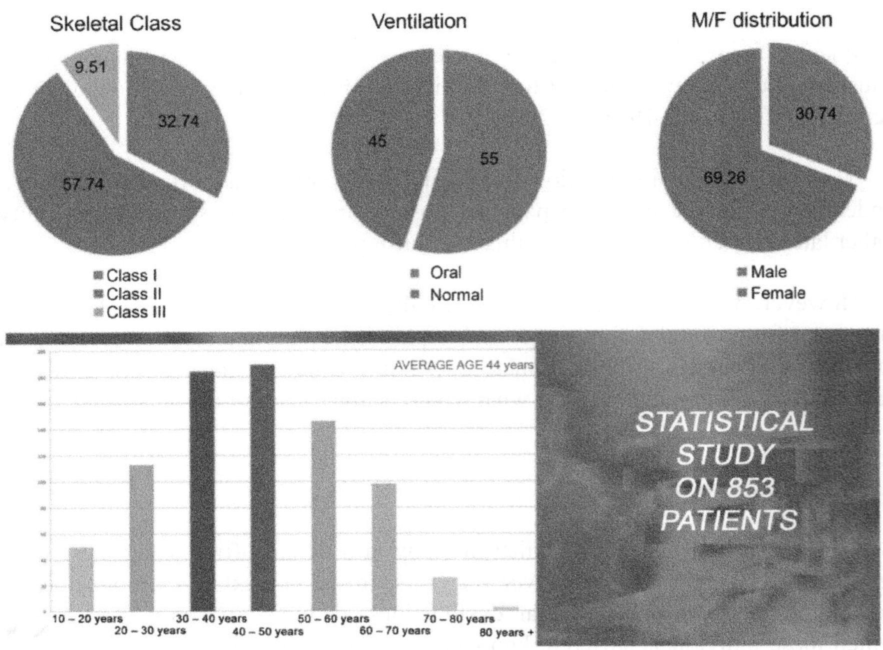

Figure 8.3. *Statistical results by skeletal class, ventilation and sex. For a color version of this figure see www.iste.co.uk/dambricourtmalasse/futureofsapiens.zip*

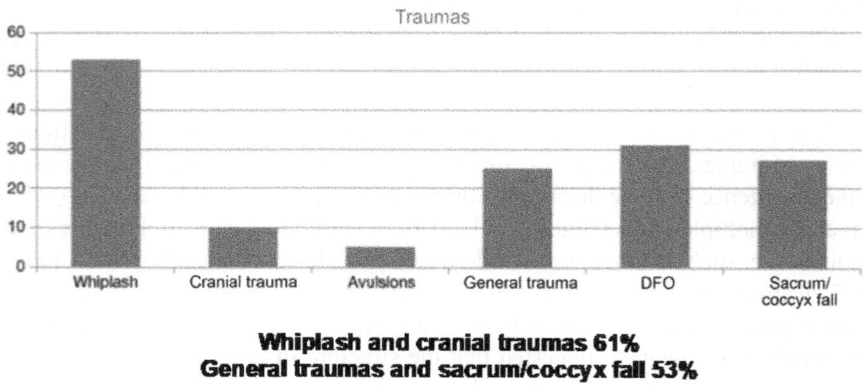

Figure 8.4. *The different traumas*

8.2.1.2. *Hypothesis on gender differences*

For our part, we put forward a more original hypothesis. Women present with more coccygeal-sacral lesions than men, linked to their role as parturients and to all the traumas of childbirth: cesarean sections, and forceps, not forgetting epidural anesthesia, which can create tearing in the dura mater. They also appear to present with more coccyx-sacrum falls than men (Clauzade and Clauzade 2006). Vertical posture is demanding and, by definition, unstable and requires the establishment of self-control systems, with the development of an increasingly complex cerebellum. Falls and traumas will be essential factors in its decompensation.

In our statistical analysis, we find a cranial etiology (cranial trauma and whiplash) in 61% of patients, and general traumas with coccygeal etiology in 53% of patients. The traumatic hypothesis needs to be retained for all patients, even though it does not explain all of the difference in male–female pathology.

Our modeling of the standing man through what we called the craniosacral-mandibular system explains perfectly through its two poles the genesis of pathologies.

This craniosacral system was borrowed from osteopaths, to which we added the mandible, which is the mobile element. It is a dural system, fluid via the cerebrospinal fluid, which unites the sacrum to the skull (Clauzade and Darraillans 1989; Clauzade and Marty 1998) (Figure 8.5).

This systemic representation allows us to understand:

– The existence of two self-organized systems: the craniomandibular and the craniosacral, which will interact with each other.

– The TMJ becomes the adapter center of this system, hence the role of cranial suture that we attribute to it.

– The trigeminal innervates the TMJ and the cranial dura maters: the *tentorium cerebelli* and the *falx cerebri*, which osteopaths consider "reciprocal tension membranes" that regulate cranial tensions. We then discover the physiological function of bruxism, which, through a clenching of the teeth, causes a trigeminal action that releases intracranial tensions. This physiological reflex will become harmful to the manducatory apparatus in the long term, and it will be the dentist's role to regulate it.

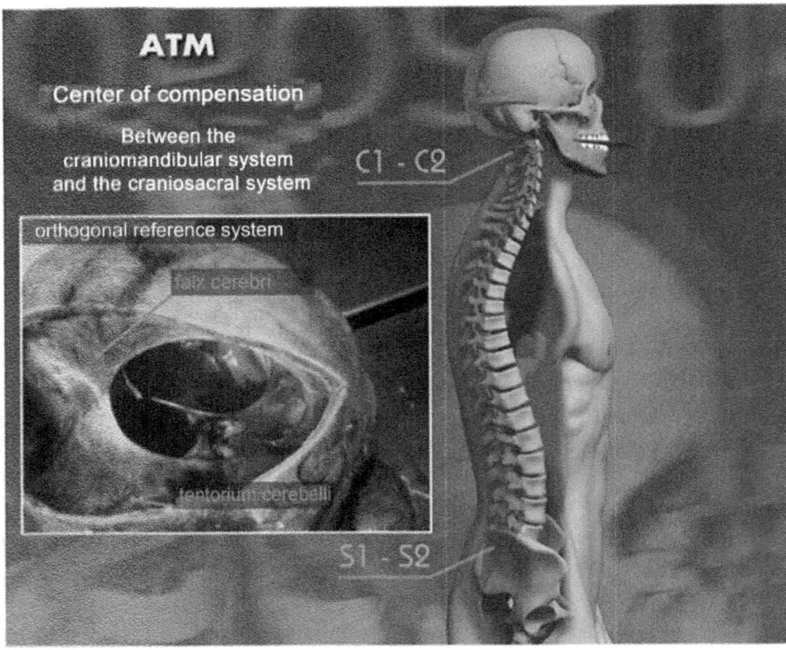

Figure 8.5. *Craniosacral-mandibular system. For a color version of this figure see www.iste.co.uk/dambricourtmalasse/futureofsapiens.zip*

8.3. Oral and clavicular ventilation

We were surprised to note that 55% of patients presented with archaic swallowing associated with a lingual dysfunction responsible for oral ventilation, which was in more than half of our patients. For the majority of dentists and orthodontists, it is the presence of a low-lying tongue that creates oral ventilation. This chicken-and-egg debate has been occurring for decades in our profession and should not exist. The first function of a newborn is breathing; the first food of a living being, long before milk, is air and oxygen. We will not dwell on the degradation of the air that we breathe owing to environmental causes, but two examples might give us cause for thought:

– Some suggest that the presence of large animals such as dinosaurs 300 million years ago was due to a 30%-higher concentration of oxygen in the air, linked to a more active photosynthesis in relation to existing forests and plants (Reichholf 1994). For the first time, the US Army reportedly observed a decrease in the size of its conscripts, while the size of individuals was steadily increasing due to improved sanitary and nutritional conditions.

– The volume of air inhaled per minute was of the order of 5 liters in the 1970s and 1980s, and we would today find values of 6 to 7 liters, which shows that to have physiological gas exchanges between CO_2 and O_2, the individual is required to inhale a greater quantity of air, which can lead to heart failure, arterial vasoconstrictions, hypocapnia and asthma (Clauzade and Clauzade 2013, 2016).

– Oral breathing is established very early in infants and is of emotional traumatic origin, signaling a loss of mother–child attachment. For Françoise Dolto, the nose serves to smell the mother, and therefore the infant breathes through its nose: in order not to lose this olfactory attachment, she suggested that children separated from their mothers at birth (premature babies, or hospitalization of the mother) keep with them a piece of cloth, an object steeped in the mother's scent (Dolto 1994). For Bullinger, oral breathing is a postural tool that the baby will use, in the absence of a tonic-postural axis, to tone and stiffen its longitudinal axis and allow the straightening of the torso, which will lead to orientation functions (Bullinger 1996). He speaks of "pneumatic tone" (Bullinger 2004). This too involves poor psychomotor development.

In our analysis, a malocclusion is a somatic marker of psychomotor dysfunction (Clauzade and Marty 1998, 2006): the reflection must extend beyond the teeth.

But another phenomenon has attracted our attention for the past 10 or so years, which is not yet detected by our profession. Thomas (2014) sounded the alarm with regard to a usual posture of which we have not considered all the consequences: the sitting posture.

In the past, we used to talk about 3x eight hours: eight hours lying in bed, eight hours standing and eight hours sitting. The sitting position devoured the rest, other than the slouching that dominates in adolescents. This position creates a relaxation of the *transversus abdominis* and a much lower effectiveness of the diaphragm, which leads to a low-intensity inhalation comparable to oral ventilation of the order of 350 ml instead of 500 ml (Clauzade and Clauzade 2013). It will promote sleep apnea, which is much talked about today. We are talking about clavicular ventilation. Campignion (2012) talks about adynamic breathing for a seated man, whereas dynamic nasal-diaphragmatic breathing can only occur in a standing man. This is a significant evolutionary finding.

Nasal breathing or nasal-diaphragmatic ventilation is fundamental for standing humans, for the acquisition of verticality, and it is a fact too often ignored in posturology, as it is only taught by podiatrists, osteopaths or ophthalmologists:

– Nasal breathing allows the development and growth of the upper maxilla; otherwise, we will witness the installation of endognathy.

– Nasal breathing helps obtain a straight head in space, aligned in the Frankfurt plane (visuo-labyrinth plane) with a cervical lordosis; in contrast, a mouth breather will present with "turtle head" (head projected forward) with a cervical kyphosis (Clauzade and Clauzade 2016) (Figure 8.6).

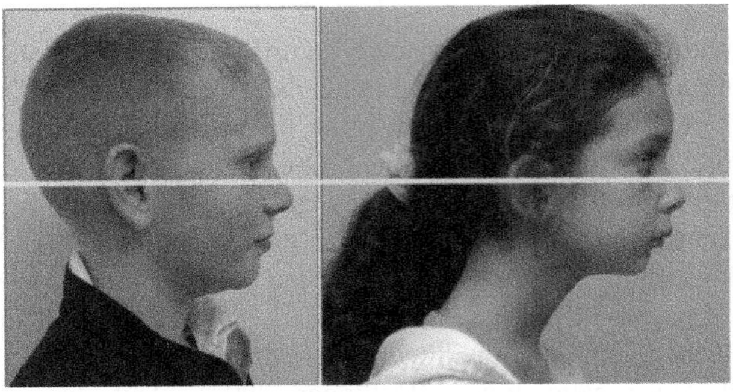

Figure 8.6. *Right side of head with nasal-diaphragmatic ventilation. Turtle head of patient breathing orally. For a color version of this figure see www.iste.co.uk/dambricourtmalasse/futureofsapiens.zip*

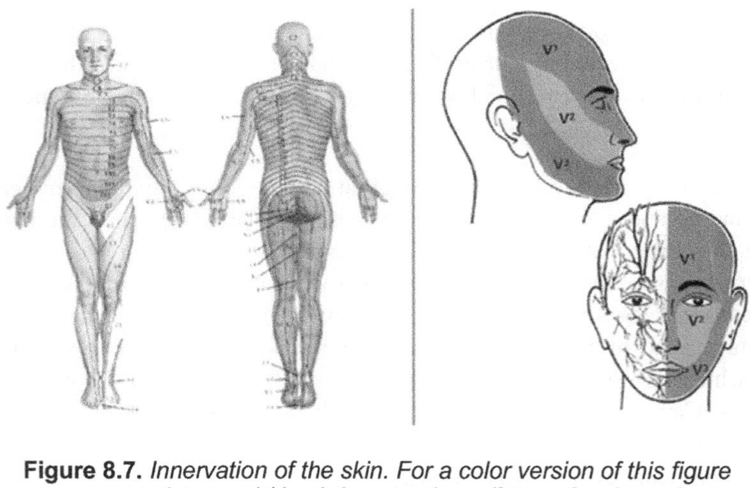

Figure 8.7. *Innervation of the skin. For a color version of this figure see www.iste.co.uk/dambricourtmalasse/futureofsapiens.zip*

Nasal breathing is a cerebral thermoregulator and cools the brain by 1.4° (Talmant 1992, 1995). We speak of encephalization, but the brain is the organ that consumes the most energy in the human body: 30% to 35% of blood glucose in adults, up to 66% of blood glucose in infants, is used for brain development and function, for an organ that represents 1/50 of the total mass of the individual

(Clauzade and Clauzade 2016). It is gigantic, and its temperature will therefore have to be regulated, because it is very sensitive to hyperthermia, hence the action of the nasal cavities, which create turbinal cooling, operated at the venous-blood level. The second source of cooling is given by the skin of the face and head, which produces perspiration and sweat. The cephalic skin is innervated by the three branches of the trigeminal: V1, V2 and V3. Let us not forget that the vertical upright position has allowed the brain to be moved away from the ground and minimized its thermal effects (Clauzade and Clauzade 2016). In this period of global warming, we suggest to our patients that they wear a hat, preferably straw (Figure 8.7).

8.4. The skeletal classes

The distribution of skeletal classes, according to the Angle classification (1899), brings us very significant evolutionary elements.

8.4.1. *Skeletal class I*

Skeletal Class I, with orthognathic typology, which represents for us an adaptive biotype, which is our evolutionary model leading to verticality, is present in 33% of cases, which is not insignificant. This shows the significance of cranial, coccygeal or general traumas in postural deficits. It is conceivable that dental occlusion is a condition that is necessary but not sufficient for postural equilibrium. Postural deficits are multifactorial, as are temporomandibular dysfunctions.

8.4.2. *Skeletal class II*

Skeletal Class II, with retrognathic typology, is by far the most represented, with 58% of cases. It should be noted that Class II is an expansionary skeletal class linked to masticatory and behavioral changes that have emerged over the past 40 years: its rapidity on a time scale should capture the attention of anthropologists. As early as 1982, Slavicek reported the preponderance of Class II patients in dental offices in Germany and Austria (Slavicek and Schadlbauer 1982). We do not have statistical data on the distribution of Class II individuals in a national population. How can this evolution be explained?

The mandible is a cranial shaper, and its function has continuously weakened, especially in great apes, where it played a role in defense, and food grinding (Clauzade and Daraillans 1989). We believe the emergence of fire and the cooking of food, agriculture and utensils for cutting and chopping food led to a reduction in its activity and contributed to the emergence of the psalidontal dental articulation of *Homo sapiens* and its orthognathic skull. Language and speech civilized this oral function, but it maintained a certain muscular tonic activity over several millennia. Modern post-industrial life has seen a reduction in the time spent on meals, meeting up and sharing, in favor of snacking, bought or ordered meals made from frozen industrialized products that are high in sugar and salt and increasingly soft or pasty in texture. Groups and the convivial have given way to isolated and depressed individuals. Individuals no longer chew; they swallow, with the time spent chewing per day estimated at 20 minutes in the 1950s and 1960s. How much is it today? Worse still, speech has disappeared in favor of the eye, text messages and screens. Vision and visual areas are taking up increasing space in cerebral processing. The result is that this mandibular hypofunction leads to a retromandibulia combined with an expansion of the prefrontal areas, a continuation of the craniofacial contraction of Dambricourt Malassé. It is the speed of evolution that raises questions (Figure 8.8).

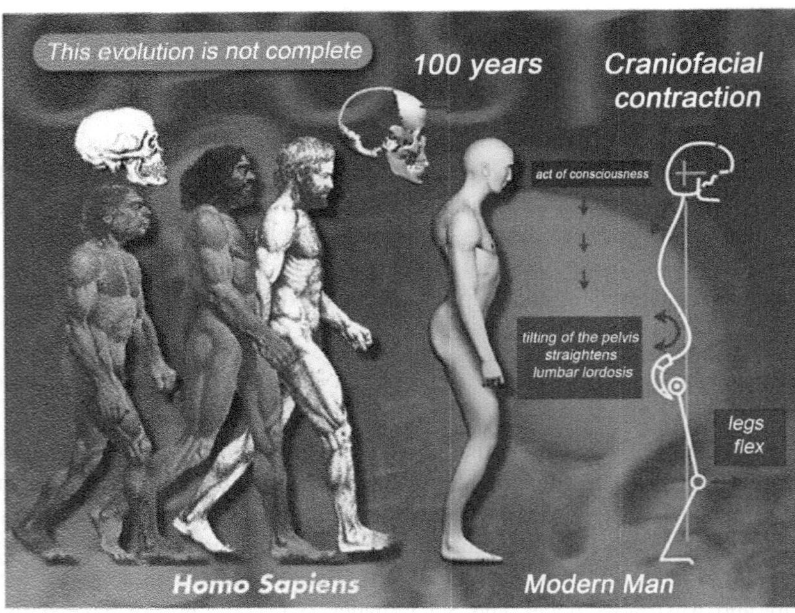

Figure 8.8. *Cranial and postural evolution. For a color version of this figure see www.iste.co.uk/dambricourtmalasse/futureofsapiens.zip*

What is important to remember is the fact that these patients have lost verticality. If we examine them in the Romberg test (Clauzade and Marty 2006; Vallier 2014;

Clauzade and Clauzade 2016), which is a postural examination, with arms outstretched in front of oneself and eyes closed, these people are generally unstable and oscillate backwards. As a result, to counter this internal posterior tilting, they adopt an eyes-open position, a compensating anterior posture that they obtain by flexing their knees and creating a lordotic curve in their pelvis. This is a compensatory biotype (Clauzade and Marty 2006; Clauzade and Clauzade 2016; Bricot 2020). We are not talking about pathology in the strict sense of the term, but about a posture that is in essence compensated and therefore more weakened in everyday life. The results of our investigation fully validate this diagnosis.

In our study, we did not differentiate between different Class IIs, with Class II 1 being a Class II with incisor infraclusion and Class II 2 with incisor supraclusion. From a medical postural perspective, Class II 2s are more deleterious due to the anterior locking they offer.

8.4.3. *Skeletal class III*

Skeletal Class III, with prognathic typology, represents only 9.5% of cases. The expansionism of Class IIs explains why Class IIIs are underrepresented in our practices. The genetic determinism that seemed to support them in Spain and Germany under the Habsburgs seems to be dissipating. Paleoanthropology has shown that verticality is acquired with *Australopithecus* and the genus *Homo*, but their body is not as centered on the Earth's axis of gravity as *Sapiens*; *Australopithecus* are in Class III and the fossil species of the genus *Homo* are labidontal (Dambricourt Malassé 2010).

In the Romberg postural examination, individuals in Class III present with anterior oscillations, as if the verticalization process were not complete, and with their eyes open, they adopt a compensating posterior posture to counter this anterior failure. Their carbocalcic typology and their "warrior" temperament make them less vulnerable to postural pain. However, they will be much more difficult to treat in a therapeutic context.

8.4.4. *Age distribution*

The average age of the patients is of the order of 44 years, but it seems that the peak of pathologies occurs between 35 and 55 years, so in the aging adult. We believe that this is the effect of gravity, which tends to compress the spine and amplifies or triggers postural issues. It is a cofactor, without being the underlying etiological cause.

8.5. Conclusion

This study, conducted based on clinical data concerning a panel of 853 patients treated for temporomandibular dysfunction, suggests to us interesting evolutionary findings:

– Nasal-diaphragmatic ventilation is an essential function of the individual that must be respected, controlled and monitored by dentists and also by doctors. Rehabilitation must be undertaken with exercises performed while standing, combined with proprioceptive rehabilitation of the other postural sensors, because it results from a psychomotor and emotional disorder. Singing and the sung voice can also serve as therapeutic means (Clauzade and Clauzade 2016). Oral breathers should be treated as soon as possible: "Shut your mouth and save your brain!", as Guye said in 1889 (Clauzade and Clauzade 2016).

– Skeletal Class II, which would be the ultimate evolution of modern man, shows a postural weakening with a compensatory anterior posture and a loss of verticality.

Is it a coincidence that we have been confronted with respiratory viruses in recent years: SARS and COVID? Is it a coincidence, if the retrognathic skull corresponds to the last geometric and physical limit of the cerebral process of craniofacial contraction evoked by Dambricourt Malassé (2022a, 2022b)?

For the moment, all these dysfunctions can be reversible, treated, provided they are diagnosed in young children, but they require adherence, the understanding of the patient and their willingness to make changes to their behavior. Environmental, societal and political issues will be much more difficult to modify. Awareness within society and the medical world is needed.

8.6. References

Angle, E. (1899). Classification of malocclusion. *Dental Cosmos*, 41, 248–264.

Bracco, P., Debernardi, C., Oiancino, M.G., Salvetti, G., Bazzichi, L. (2010). Evaluation of the stomatognathic system in patients with rhumatoid arthritis according to the research diagnostic criteria for temporomandibular disorders. *Cranio.*, 28(3), 181–186.

Bricot, B. (2020). *La reprogrammation posturale*, 2nd edition. Sauramps Médical, Montpellier.

Bullinger, A. (1996). Le rôle des flux sensoriels dans le développement tonico-postural du nourrisson. *Motricité cérébrale*, 17, 21–32.

Bullinger, A. (2004). *Le développement sensori-moteur de l'enfant et ses avatars*. Érès, Toulouse.

Campignion, P. (2012). *Respir-actions*. Frison-Roche, Paris.

Carpentier, P. and Felizardo, P. (2015). Ostéoarthrose, ostéoarthrite, arthrite juvénile idiopathique. In *Troubles musculo-squelettiques de l'appareil manducateur*, Fleiter, B., de Jaegher, P., Fougeront, N.T. (eds). Quintessence international, Paris, 107–120.

Clauzade, M. and Clauzade, N. (2013). *Dents et performances sportives*. Chiron, Paris.

Clauzade, M. and Clauzade, N. (2016). *Le capteur mandibulo-spinal*. SEOO, Perpignan.

Clauzade, M. and Darraillans, B. (1989). *Le concept ostéopathique de l'occlusion*. SEOO, Perpignan.

Clauzade, M. and Darraillans, B. (1992). *L'homme, le crâne, les dents*. SEOO, Perpignan.

Clauzade, M. and Marty, J.P. (1998). *Orthoposturodontie*. SEOO, Perpignan.

Clauzade, M. and Marty, J.P. (2006). *Orthoposturodontie 2*. SEOO, Perpignan.

Damasio, A. (1999). *Le sentiment même de moi*. Odile Jacob, Paris.

Dambricourt Malassé, A. (2010). The relationships between occlusion and posture in the hominid lineage, implications for the transition between Mesolithic and Neolithic populations. *International Journal of Modern Anthropology*, 3, 14–6.

Dambricourt Malassé, A. (2022a). *Embryogeny and Phylogeny of the Human Posture 1: A New Glance at the Future of our Species*. ISTE Ltd, London and Wiley, Hoboken, NJ.

Dambricourt Malassé, A. (2022b). *Embryogeny and Phylogeny of the Human Posture 2: A New Glance at the Future of our Species*. ISTE Ltd, London and Wiley, Hoboken, NJ.

Deshayes, M.J. (1986). *Croissance cranio-faciale et orthodontie : rapports de biodynamique crânienne*. Masson, Paris.

Deshayes, M.J. (1991). Reconsidération de la croissance maxillo-faciale au cours du développement et de l'évolution, conséquences en orthopédie dento-maxillo-faciale. *Rev. Orthodento. Fac.*, 25, 353–365.

Dolto, F. (1994). *Les étapes majeures de l'enfance*. Gallimard, Paris.

Hadjouis, D. (2021). *The Skull of Quadruped and Bipedal Vertebrates: Variations, Abnormalities and Joint Pathologies*. ISTE Ltd, London and Wiley, Hoboken, NJ.

Ito, M. (2001). Cerebellar long-term depression: Characterization, signal transduction and functional roles. *Physiol. Rev.*, 81(3), 1143–1195.

Le Double, A.F. (1903). *Traité des variations des os du crâne de l'homme et leur signification du point de vue de l'anthropologie zoologique*. Vigot, Paris.

Lin, Y.C., Msu, M.L., Yang, Y.S., Liang, T., Chou, S.L., Ling, H.Y. (2007). Temporomandibular joint disorders in patients with rhumatoid arthritis. *J. Chin. Med. Assoc.*, 70, 527–534.

Reichholf, J. (1994). *L'émancipation de la vie*. Flammarion, Paris.

Slavicek, R. and Schadlbauer, E. (1982). Étude et comparaison des valeurs céphalométriques en Autriche et Allemagne. *Rev. Orthodento. Fac.*, 16, 417–471.

Strazielle, C. (1996). Étude neuroanatomique de voies et sites trigéminaux et non trigéminaux impliqués dans le système manducateur à l'aide de marqueurs vitaux en fluorescence et histochimie chez le rat. PhD Thesis, Université de Nancy, Nancy.

Stricker, M. and Raphael, B. (1993). *Croissance crânio-faciale*. Morphos, Reims.

Talmant, J. (1992). Du rôle des fosses nasales dans la thermorégulation cérébrale. *Rev. Orthodento. Fac.* 26, 51–59.

Talmant, J. (1995). Ventilation et mécanique des tissus mous faciaux : en guise de préambule. *Rev. Orthodento. Fac.*, 29, 175–188.

Thomas, H. (2014). *Posture et respiration. Le poids et le coût de la vie*. Spire, Paris.

Vallier, G. (2014). *Traité de posturologie. Clinique et thérapeutique*. Posturopole, Perpignan.

9

Homo sapiens and Verticality: The Contribution of Research in Space Medicine and Posturology

9.1. Introduction

Research in space medicine and posturology over the past three decades has led to major advances in our knowledge of the physiology of antigravity control in *Homo sapiens*. The experience of weightlessness has highlighted the crucial role of gravity in our evolution and our physiology. Thus, all physiological systems are impacted by this extreme environment of microgravity (μG) (Buckey 2006; Gunga 2015; Custaud et al. 2020). Terrestrial gravity is thus a major element in our evolution on Earth. It is what shapes our bone growth and, more generally, the mass of our bony skeleton. Terrestrial gravity is also responsible for our muscle development; it plays a major role in fluid distributions and their regulation in our bodies. Gravity constitutes a real reference point that has partly guided the phylogenetic evolution of our species. This evolution, resulting in an exclusive bipedalism that is biomechanically unstable, the maintenance and control of which pose numerous problems to be solved, potentially leading to chronic and recurrent neuro-musculoskeletal disorders such as lumbago[1]. Within this public-health context, rehabilitation medicine and manual therapies have notably developed, including osteopathy, which the World Health Organization (WHO) defines as a complementary medicine that focuses on the diagnosis and treatment of functional

Chapter written by Denis DUCOMMUN and Loïc TREFFEL.

1 WHO guideline for non-surgical management of chronic primary low back pain in adults in primary and community care settings. World Health Organization, Geneva, 2023. License: CC BY-NC-SA 3.0 IGO. ISBN 978-92-4-008179-6.

somatic disorders[2]. Osteopathic manual medicine aims to increasingly pervade allopathic medicine, serving person-centered health, as recommended by the WHO (Von Schoen-Angerer et al. 2023).

In the recent context of space exploration, particularly of Mars, increasingly long flights are being envisaged. Numerous elements of physiological adaptation to prolonged exposure to microgravity are still to be investigated. A round trip to Mars represents about 500 days in space. Immediate as well as long-term questions relating to adaptations from *Homo sapiens in terra* to *Homo sapiens in spatio* remain. These questions can be developed on different levels, including the anthropological and philosophical. At the musculoskeletal level, many questions and needs persist in the idea of going further and for longer into space (Liphardt et al. 2023). Research in space medicine and clinical posturology helps shed additional light on the adaptation of *Homo sapiens* to verticality. These questions are asked from a morphological standpoint: what will *Homo sapiens* look like in the future? What are the implications for current practice, both in rehabilitation medicine and in manual therapy? This is precisely the purpose of this chapter, which will firstly expand on the control of the upright posture of *Homo sapiens*, the effects of microgravity on human physiology, and the avenues of reflection and application in current clinical practice, in rehabilitation and in osteopathic manual medicine.

9.2. Control of upright posture in *Homo sapiens*

9.2.1. *Construction of the representation of verticality*

The works of Gilles Clément have helped bring to light evidence of the existence of an internal model of verticality (Clément 2003; Moore et al. 2003; Clément and Wood 2014). Indeed, astronauts placed in a centrifuge on Earth and subjected to a centrifugal acceleration of 1 g perceive a 45° body tilt due to the gravitational acceleration (sum of the gravitational and centrifugal acceleration vectors). In weightlessness, the same experiment, taking into account the disappearance of gravitational acceleration, should result in a perceived 90° body tilt. However, this result is only obtained after about 15 days of weightlessness, evidence of the existence of an internal representation of verticality passing through the body axis (Barra et al. 2010).

The construction of the representation of verticality is conducted through a model integrating somatosensory, vestibular and visual information that comes and mixes with our internal model of verticality (Dakin and Rosenberg 2018). This

2 Benchmark for training in osteopathy, WHO, 2010. WHO Headquarters Team.

construction will serve as a basis for the development of our spatial representation and will allow the implementation of a relevant motor control in a gravity situation, thanks to the implementation of high-quality anticipatory postural adjustments. Anticipatory postural adjustments (APAs) designate the mechanisms by which the human body anticipates and prepares muscle responses, before a disruption of balance even occurs. These adjustments may involve early activation of the stabilizing muscles of the trunk, limbs and neck, as well as a modification in body weight distribution, in order to maintain a stable posture. Thus, when you extend your arms out in front of you, you shift your center of gravity forward. About 200 milliseconds before your deltoids contract, the first muscles to contract are the solei. The APAs, which are entirely non-conscious, are the basis of our adaptation and our motor performance. They are built and improved through the repetition of active scenarios, a largely procedural learning process (Bouisset and Zattara 1987).

The functioning of this representation of the verticality model is still in the process of being validated. However, we can cite the works of Ksander N. De Winkel on the "subjective visual vertical", which, through a Bayesian model mixing multisensory inputs and the internal model, results in a probabilistic construct of the vertical (De Winkel et al. 2018, 2022). However, "recent discoveries further show that the weighting of the respective sensory signals is proportional to their reliability, which corresponds to a Bayesian interpretation of a sum of vectors. However, violations of this model have also been reported, suggesting that the CNS may rely on a single sensory system or choose to process sensory signals based on inferred causality from the signal". The concept of vicariance developed by Berthoz, therefore, still appears to be relevant (Berthoz et al. 1979). Depending on the conditions, the weight of the different sensory inputs in the equation of the postural model of verticality will be adjusted; this adjustment will be done in an idiosyncratic manner.

The three main somatosensory inputs correspond to the three postural reference frames generally studied and explored in posturology: the geocentric postural reference frame for the vestibule, the egocentric postural reference frame for the somatosensory system and the allocentric postural reference frame for vision. There exists a famous experiment conducted by Richard Fitzpatrick's team, designed to measure the sensitivity of the different postural reference frames to antero-posterior tilt, as a function of angular velocity, in subjects in an upright antigravity posture (Fitzpatrick and McCloskey 1994). The implementation of different experimental paradigms allowed each of the postural reference frames to be isolated. The limitation of this study was that the tilt sensation was declared orally by the participant. This study highlighted that muscle proprioception alone is as sensitive as the three reference frames combined. The least sensitive reference frame in this situation is, surprisingly, the geocentric reference frame. These three coupled reference systems form a genuine inertial unit. The vestibular system will capture accelerations during movements; the proprioceptive system, changes in length (muscle stretching); and the visual system, movement speeds of the visual scene (peripheral retina).

9.2.2. *The geocentric reference frame*

The vestibular system is composed of two main parts: the otolithic organs and the semicircular canals. The otolithic organs are specialized in capturing translational (linear) accelerations, while the semicircular canals are specialized in angular rotational accelerations. Together, the semicircular canals and otolithic organs provide information on the movements permitted in three-dimensional space.

In the case of vestibular information, a good example of their role is the deviation of the body axis in the frontal plane and the tonic asymmetry encountered during unilateral surgical destruction of a vestibular nerve. The patient will initially tilt in the frontal plane on the side of the lesion due to the weight of the vestibular error in the construct of verticality, but over time and readaptation, this frontal decentralization will disappear thanks to the modification of the gain of the different postural inputs (McDonnell and Hillier 2015). However, several years later, if the person is placed in a condition that requires significant gain in vestibular input (eyes closed on an unstable plane), the positional and tonic defect will reappear, demonstrating that purely vestibular information in the construct remains problematic.

In a standing subject, in a static or dynamic condition, the positioning of the head in flexion or extension modifying the positioning of the vestibular system induces a decrease in the quality of the postural control (Buckley et al. 2005; Paloski et al. 2006).

Furthermore, the sensitivity of vestibular perception is likely to be disrupted quite easily and in a relatively long-term manner by simple passive turning maneuvers, as highlighted by Fitzpatrick and Watson (Fitzpatrick and Watson 2015). When managing a patient presenting changes in position on an examination table, caution should therefore be exercised as to the effect measured, immediately after treatment or manipulation, on tone and antigravity control.

9.2.3. *Egocentric reference frame: proprioception*

The term "proprioception" was coined by Sherrington in the early 20th century, declaring: "In muscular receptivity we see the body itself acting as a *stimulus* to its own receptors – the proprioceptors" (Sherrington 1907). It is now accepted that proprioceptive signals comprise peripheral inputs from muscle spindles, Golgi tendon organs, skin and joint receptors, and central inputs originating from efferent motor controls. Proprioception can therefore be defined as awareness of the mechanical (static and dynamic) and spatial state of the body and body segments. Two types of proprioception are defined: "low-level" proprioception, which refers to a single modality, such as perceiving at what moment a body segment is moving, and "high-level" proprioception, which refers to several modalities, such as determining the length of a hidden object held between the thumb and index finger by comparing it to straight lines of different lengths (involving both a muscular perception modality and a visual modality). In static upright posture, the role played by podal plantar information is considered negligible (Hochet Russell 2016).

The contributions of the experimental exploration of this postural reference frame are of great importance for manual therapists and rehabilitation therapists, since they intervene and build their reasoning mainly on the musculoskeletal anatomical system as they provide their care (Leach et al. 2019; Hidalgo et al. 2024).

To highlight the power of proprioceptive input in the sensory mix, we can refer to the works of James R. Lackner on the effect of muscle vibrations, which shows that muscle proprioception modifies visual perception by creating illusions of movement (Lackner and Dizio 2005). Let us recall that the vibration of a muscle, at a frequency of 80 Hz, gives an illusion of the stretching of this muscle. For example, in the case of vibration of the biceps brachii, it will give the subject an illusion of the extension of the forearms. If we place two light-emitting diodes on the index fingers of a subject placed in the dark, in this vibration condition, they will declare that they visually perceive a distancing of the diodes, yet there is no movement of the forearms.

Proprioception plays a major role in the mix of postural afferents. For example, it could be thought that people suffering from chronic lumbago, presenting with a deficit of lumbar proprioceptive afferents in maintaining antigravity posture, would substitute these elements with an overuse of visual or vestibular afferents to stabilize themselves. However, this is not the case: we witness a reorganization of segmental proprioceptive afferents to compensate for the deficit (Mohammadi et al. 2021).

The role of proprioception cannot be evoked without referencing the role of oculomotor vergence in posture control. Indeed, the external oculomotor muscles present a wealth of neuromuscular spindles and Golgi organs that are out of the ordinary (Wasicky et al. 2000). Experiments with oculomotor muscle vibrations and their effects on orthostatic posture have been well documented, notably by Roll and colleagues (Roll et al. 1989; Bertrand-Charette et al. 2023). The vibration of the superior rectus muscles causes the subject to tilt forward; the vibration of the *rectus lateralis* muscle of the left eye and the *rectus medialis* muscle of the right eye causes the subject to tilt to the left. Convergence, involving contraction of the *rectus medialis* muscles (and therefore a stretching of the *rectus lateralis* muscles), resulting in optical axes getting closer and an increase in the eye vergence angle, allows the fusion of the two retinal images into a single image (this is fusional vergence). The greater this angle of convergence, the more stretched the lateral oculomotor muscles, and the more stable the subject is in upright posture. This assertion is verified only in conditions of binocular vision (Lê and Kapoula 2006). It is therefore important to take this element into account when performing posturographic measurements; they require a standardization of the distance to the visual cue in order to be valid. The exploration of oculomotor imbalances is therefore an absolute necessity whenever we are researching postural control and its disturbances in daily clinical practice.

9.2.4. *The allosteric reference frame*

Vision is the dominant sensory input in maintaining and controlling upright posture, as well as in our movements within our environment (Yelnik 2021). In this section, we will focus on retinal vision, the role of visual convergence having been discussed in the previous section. The peripheral retina plays a crucial role in the perception of movements of the head or the visual scene. By perceiving the speed and direction of the optical flow during postural oscillations or movements, retinal vision makes it possible to finely adjust our balance, our movements and our spatial navigation. Numerous experiments manipulating visual flow prove the power of this sensory input, the pioneer being J. J. Gibson (Ogle 1951; Gibson et al. 1955; Gibson 2014). A concrete example, one that everyone has been able to feel, is the phenomenon of vection, i.e. the subjective experience of moving. This illusion of movement occurs in the opposite direction to visual stimulation (Dichgans and Brandt 1978). You are sitting on a train or bus; when a train or bus adjacent to yours starts moving, you perceive the illusion of moving in the opposite direction and you will apply a retroactive postural adjustment in order to correct your illusory imbalance. Experiments on oscillatory movements, superimposed on a visual simulation of forward movement (using virtual reality, for example), highlight an improvement not only in the feeling of vection (Palmisano et al. 2000; Kim et al. 2012) but also in the sensation of walking or running (Bubka and Bonato 2010). As we will see below, the decrease in the quality of this input during aging is one of the major factors in the risk of falling.

9.2.5. *Maturation of the postural system*

The evolution of motor and postural development is closely associated with the age of the child and the myelination of the different areas of the brain. In early childhood (0 to 2 years of age), myelination is concentrated mainly on the brain-stem and primitive-brain regions. The first few months are characterized by a rapid improvement in basic postural control, including the ability to hold the head upright and sit with support. Primitive postural reflexes, such as the straightening reflex and the tonic reflex of the neck, are present from birth and gradually improve. During late infancy (2 to 5 years of age), myelination progresses in the limbic-brain regions. Children acquire basic motor skills such as walking, running and jumping, thanks to improved motor coordination and more-mature postural responses. They also develop a better ability to adjust their posture in response to the demands of the environment and activities. Then, during middle childhood (6 to 11 years of age), myelination continues in associative cortical regions, including supplementary motor areas and regions involved in higher cognitive functions such as the prefrontal cortex. Children acquire more advanced motor skills, such as hand-eye coordination and object manipulation. Their postural control becomes more sophisticated, with an increased ability to maintain balance in static and dynamic situations. As shown in the study by R. Cuisinier (Cuisinier et al. 2011), during middle childhood, for example, a disruption of the proprioceptive system by a vibration of the Achilles tendons would cause a much more marked postural imbalance than in adults, whose postural system is mature and capable of modulating the afferents of the postural sensors. Particular caution should therefore be exercised when installing disruptive external elements such as orthodontic devices or foot orthoses in children. During adolescence (12 to 18 years of age), myelination of the prefrontal cortex continues, contributing to the development of executive functions. Adolescents develop more advanced fine and gross motor skills, as well as better body awareness and improved coordination. Their postural control becomes more stable and adaptable. It would appear that the improvement in the performance of our antigravity stabilization system improves further in adulthood, since the quality of stability descriptors measured in stabilometry improves up to the age of 35 (Rossato et al. 2013).

9.2.6. *Aging of the postural and motor system*

Aging of the vestibular system can have a significant impact on postural stability and perception of movement. With age, there is a decrease in otolith density and an alteration in their structure, which can lead to a decrease in their sensitivity to gravity and linear movements. The sensory cells that transmit signals from the otoliths to the central nervous system will undergo alterations that result in a decrease in the precision of the transmission of information regarding orientation and movements of the head. We see a reduction in the vestibulo-ocular reflexes, which enable a stabilizing of the gaze during head movements, especially during rapid movements of the head.

Aging of the proprioceptive system is a complex process influenced by several underlying mechanisms. Sensory receptors such as neuromuscular spindles, Golgi tendon organs and joint receptors undergo age-related changes. These alterations may include a decrease in the number of receptors, a decrease in their sensitivity, and changes in their structure and operation. Aging will also lead to changes in the central nervous system, including decreased neural plasticity. This decrease in neural plasticity can affect the brain‘s ability to interpret and process proprioceptive information efficiently. Studies have shown that cortical areas involved in proprioceptive perception can undergo structural and functional changes with aging, which can affect the way in which this information is processed and integrated in the brain (Tulimieri and Semrau 2023).

Aging of the visual function can influence posture and perception of movement through several mechanisms. It is common to observe a decrease in visual acuity due to changes in ocular structures such as the lens, cornea and retina. This decrease will impair the ability to perceive environmental details, impoverishing the texture of the visual scene and therefore reference points. A decrease in stereoscopic vision, which allows depth perception using both eyes, will make estimating distances more difficult, thus influencing posture and movements. The visual field will shrink, resulting in reduced peripheral vision. All these elements will contribute to an alteration in eye-body coordination.

9.3. Effects of microgravity on human physiology: deconditioning

Exposure to real or simulated microgravity (weightlessness or μG) involves multiple adaptive physiological reactions (Buckey 2006; De Abreu et al. 2017; Custaud et al. 2020). A cascade of reactions is triggered during a spaceflight, affecting all the systems of living organisms and humans a fortiori. Thus, the continuous unloading in space affects the vascular system and energy and endocrine metabolism, as well as the musculoskeletal system, such as the spine and its paravertebral muscles, as well as the various somatosensory systems. This general disadaptation to gravitational stress is called deconditioning; it goes as far as impacting our DNA, with numerous telomeric reactions, which have been demonstrated, for example, by the comparative monitoring of twins (the brothers Scott and Mark Kelly, for NASA), one staying on Earth and the other aboard the ISS for one year (Garrett-Bakelman et al. 2019). All these reactions observed in flight are physiological, in order to adapt humans to the extreme environment of space (Gunga 2015).

Being in space represents exposure to numerous different constraints, such as solar radiation (which would involve significant dermatological lesions), microgravity and even weightlessness and also prolonged confinement. Thus, going to Mars would represent about 250 days of confinement in the spacecraft. Confinement alone has significant effects on the body, notably with an increase in cortisol secretion and C-reactive protein concentration (Yuan et al. 2019).

9.3.1. *Description of several systems affected by exposure to microgravity*

The cardiovascular, respiratory, musculoskeletal, neuroendocrine and neurovestibular systems are all affected by microgravity.

9.3.1.1. *Vascular deconditioning*

Going into space modifies the distribution of fluids in the body. Thus, the absence of gravity triggers a reaction of fluid influx toward the head, known as cephalic fluid shift. This new distribution leads within the first 48 hours to intracranial hypertension (ICH), notably responsible for the “space sickness” that can be observed in astronauts. In response, diuresis will be increased and will lead to hypovolemia. Blood pressure decreases and will therefore result in orthostatic hypotension upon returning to weight bearing. This management of orthostatism is very well studied by scientists. To mimic the action of gravity on fluids, Lower Body Negative Pressure (LBNP) exists, which aims to create a vacuum in the lower part of the body. Scientists are also testing subjects’ ability to tolerate orthostatism up to the point of syncope, with a table of progressive verticalization, called Tilt. More generally, on the vascular level, a spaceflight contributes to a decrease in circulating blood volume (hypovolemia), a decrease in mean systolic arterial pressure and left ventricular mass, and stroke volume decreases immediately after flight, as does vascular resistance (Shen and Frishman 2019; Afshinnekoo et al. 2021). The arterial wall is modified, as it is during aging, where a thickening of the intima-media space at the carotid artery can be observed, indicating an increase in arterial stiffness (Hughson et al. 2016).

9.3.1.2. *Vertebral deconditioning*

Spinal statics, i.e. the shape of the spine, finds itself modified by the absence of gravity or prolonged unloading. Certain models have highlighted the relationship between lumbar load/unload and spinopelvic morphological static (Bassani et al.

2019). In a way, it is the whole "spinal health" that is affected by real or simulated exposure to microgravity. Thus, the various scientific studies show an elongation of the spine, a decrease in lumbar lordosis, more herniated discs and amyotrophy of the paravertebral muscles (Belavy et al. 2016; Green and Scott 2017) (Figure 9.1).

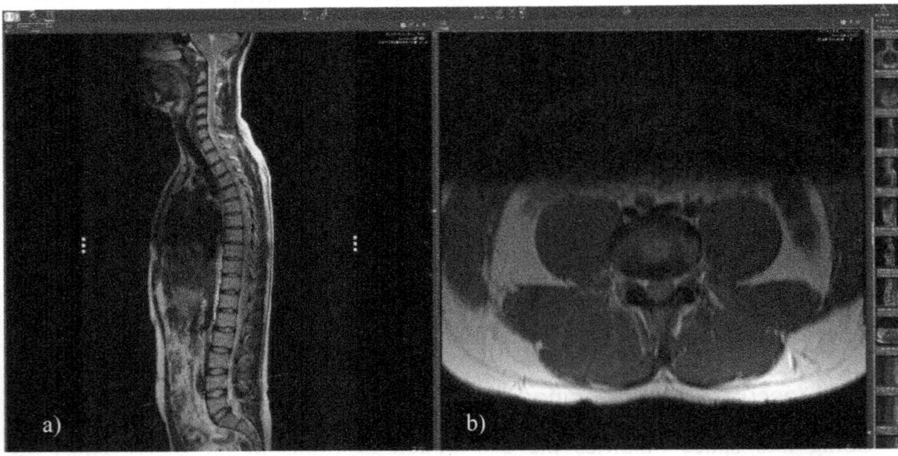

Figure 9.1. *Axial (a) and transverse (b) MRI cross-sections*

9.3.1.3. *Elongation of the spine*

The cross-sections shown in Figure 9.1 allow the spine size to be measured, from the occipital bone to the base of the sacrum. After exposure to microgravity, whether real or simulated, studies show an overall vertebral elongation, more pronounced at the lumbar level (Treffel et al. 2016; Robin et al. 2022). The cross-section notably allows the measurement of a cross-sectional area of the iliopsoas and paravertebral erector spinae muscles (*quadratus lumborum*, multifidus, longissimus and iliocostalis).

9.3.1.4. *Swelling of in vitro diagnostics volume and modification of physiological spinal curvatures*

The lengthening of the spine is coupled with an overall erasure of physiological curvatures (Figure 9.2). In this case, lumbar lordosis would decrease by approximately 8% (Treffel et al. 2016; McNamara et al. 2019).

Paradoxically to other systems in the human organism that undergo accelerated aging, the spine, in particular the IVD, does not seem to be adversely affected by this impact, notably with swelling and an increase in its volume. However, more herniated discs are observed in astronauts upon returning to Earth.

9.3.1.5. *Muscular deconditioning*

Amyotrophy of 5 to 15% and a change in the muscle typology are observed after a stay in real or simulated microgravity. Thus, global muscle atrophy of the type-I and type-II fibers is observed (Chang et al. 2016; Demangel et al. 2017; McNamara et al. 2019).

Type-I fibers, which engage the aerobic system and are generally to be found in deep muscles, often described as postural muscles, are particularly affected by a stay in real or simulated microgravity. To be cited in particular are the multifidus and soleus muscles.

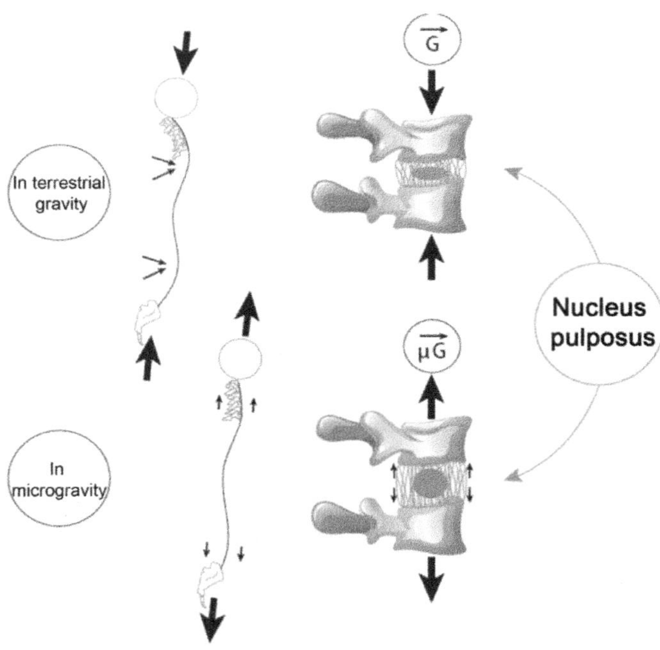

Figure 9.2. *Schematic representation of vertebral elongation in microgravity, in comparison with gravitational stress (source: (Custaud et al. 2020), L'humain et l'espace ["Humans and Space"], CNES, figure Focus 1-A.P177 by Loïc Treffel). For a color version of this figure see www.iste.co.uk/dambricourtmalasse/futureofsapiens.zip*

Fatty infiltration of the paravertebral muscles is also observed after a spaceflight, indicating muscle deconditioning similar to what would be observed in aging or in a period of severe inactivity (Greene et al. 2022).

Muscle deconditioning not only impacts motor skills but also has a systemic effect through the actions of myokines on the different systems. Myokines are defined as cytokines and other peptides produced, expressed and released by muscle fibers and exerting autocrine, paracrine or endocrine effects. Myokines ensure communication between the muscles and other organs, including the brain, adipose tissue, bones, liver, intestine, pancreas, vascular bed and skin, as well as within the muscle itself (Severinsen and Pedersen 2020).

These morphological changes in the spine, adjacent muscles and at the cardiovascular level are also concomitant with somatosensory and neurovestibular deconditioning. To combat these effects, various countermeasures are deployed by medical and scientific teams.

9.4. Countermeasures to deconditioning

Figure 9.3. *Cover illustration of the book from the International Academy of Astronautics. The image illustrates the Advanced Resistive Exercise Device (ARED) in the US segment (a) and the treadmill in the Russian segment (b) of the International Space Station (NASA) (Norsk et al. 2021). For a color version of this figure see www.iste.co.uk/dambricourtmalasse/futureofsapiens.zip*

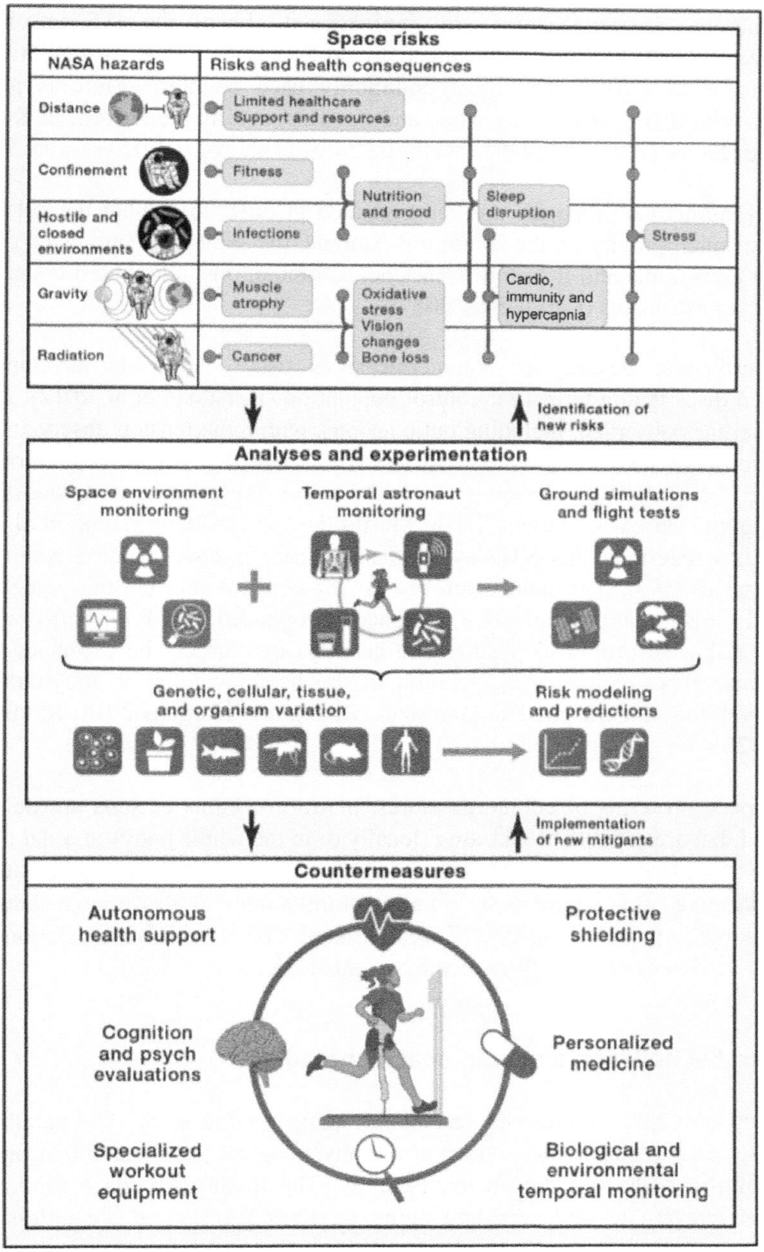

Figure 9.4. *Synthesis of the effects of microgravity on human physiology and countermeasures developed (Afshinnekoo et al. 2021). For a color version of this figure see www.iste.co.uk/dambricourtmalasse/futureofsapiens.zip*

Numerous countermeasures are implemented aboard the ISS in order to minimize the deleterious effects linked to prolonged exposure to weightlessness (Winnard et al. 2017). The most commonly used countermeasure is physical exercise with different characteristics and devices such as the Advanced Resistive Exercise Device (ARED) and the treadmill (Loehr et al. 2011, 2015) (Figure 9.3).

Also, numerous proposals are being studied in order to combat the deleterious effects of microgravity on the spine; for example, the skinsuit, a one-piece garment creating constraints and therefore somesthetic stimulations, allowing a better return to Earth for astronauts (Carvil et al. 2017).

Nevertheless, despite the countermeasures used, astronauts develop more herniated discs than a terrestrial control population (Kerstman et al. 2012). Lesions are sometimes observed, including optic lesions, with hypertension observed around the optic nerve sheath (II), with what is known as Visual Impairment Intracranial Pressure (VIIP). VIIP syndrome is one of the major risks identified in long-duration spaceflight (Zhang and Hargens 2018; Martin Paez et al. 2020; Macias et al. 2021). NASA has redefined this VIIP syndrome as Spaceflight-Associated Neuroocular Syndrome (SANS). The data examined so far confirm that chronic and slightly elevated intracranial pressure (ICP) in space (as opposed to ICP on Earth, which is more variable according to posture and activity) can largely be explained by the hydrostatic pressure loss and alteration in the hemodynamics in the intracranial circulation and cerebrospinal-fluid system (Zhang and Hargens 2018; Martin Paez et al. 2020).

In space, in terms of countermeasures, platforms, boots or suits are developed and used that are capable of vibrating (locally or in the whole body) in order to try to recreate somesthetic *stimuli* present on Earth and thus fight in part against this deconditioning. This whole-body vibration stimulation can also be associated with or compared to resistive exercise (Zuccarelli et al. 2023), which is a solution that is partially sufficient but widely studied by scientists (Kenny et al. 2017).

9.5. Parallel between a stay in space and aging

Going into space represents accelerated aging for the body. The parallel with aging is clearly identified by studies of the physiological and metabolic adaptations of the human body to a stay in microgravity. The findings of space medicine are naturally relevant to understanding aging on Earth. A stay in weightlessness is similar to a period of severe inactivity, with all the metabolic and cardiovascular adaptations that can also be observed in elderly individuals or during prolonged bed rest. Examples include the study of osteoporosis and the mechanisms of bone demineralization observed after a spaceflight. Space-medicine scientists regularly draw a parallel between the observed results and their application on Earth. This is evidenced by the evocative titles of the works of the American teams, such as Hides et al. (2017), "Parallels between Astronauts and Terrestrial Patients – Taking Physiotherapy Rehabilitation 'To Infinity and Beyond'" or indeed the works of Joan Vernikos of NASA: "Space, Gravity and the Physiology of Aging: Parallel or Convergent Disciplines?" (Vernikos and Schneider 2010).

The need for post-flight reconditioning is a necessity and is addressed by specific programs (Kerstman and Scheuring 2011; Petersen et al. 2017; Stokes et al. 2017). The goal is to avoid, or at least minimize, changes and risks of injury after a long spaceflight (Winnard et al. 2017). Regarding the spine, it is, for example, a matter of reducing the risk of astronauts developing a herniated disc when they return to weight bearing.

9.6. Spaceflight and disruptions of postural inputs

In space, vestibular information is almost nil, with inactivation of the otoliths during movements. Postural deconditioning is therefore significant after a stay in weightlessness. All postural inputs are involved in hypostimulation or even an absence of stimulations. As a result, aboard the ISS, for example, a convention is established to indicate an "up/down" visual cue in order to provide astronauts with spatial reference points. More generally, astronauts often describe their return to Earth as a "sensory slap" involving smelling odors again, the weight of holding their head up, the wind on their skin and auditory stimuli. An interview with Thomas Pesquet, which can be found in Loïc Treffel's doctoral thesis, recounts this feeling experienced by the French astronaut upon returning from his first Proxima mission, where he spent 6 months aboard the ISS (Treffel 2017, p. 18), notably the effort of holding his head up when he returned to Earth. Our earthly body is shaped by gravity. If only from a musculoskeletal point of view, the latter is a dynamic system designed to combat gravitational force and allow locomotion. The upright posture is therefore anti-gravitational par excellence. This postural balance is significantly affected by a stay in weightlessness. Subjects feel "disoriented" because all gravity-related reference points are disrupted or even absent. Cognitive dissonance between visual, kinesthetic and vestibular information is at the origin of these disorders. Solutions such as motor imagery (mentalization of motor sequences) can prove an interesting countermeasure in order to continue stimulating the same brain areas as during terrestrial locomotion (Saruco et al. 2017; Oh and Choi 2021), as well as the development of virtual-reality technologies, already tested aboard the ISS, during the latest missions. Thus, augmented virtual reality coupled with haptic robots is currently being tested aboard the ISS in order to enhance spatial stimulations and maintain a form of adaptation to the abundance of stimuli of the Earth's environment. This includes the SPRINT protocol using the Multi-Mode Exercise Device (M-MED). Virtual-reality training would notably help reduce the "space sickness" observed in astronauts (Chen et al. 2017), while positively impacting physiological and psychological well-being in microgravity (Keller et al. 2022).

Moreover, structural changes in the vestibular system are observed after a flight in weightlessness (Carriot et al. 2021). The phenomena observed after a spaceflight indicate a modified sensory experience that can go as far as a painful experience (Sauer et al. 2023). Physiological modifications, functional changes and their manifestations are thus multiple and systemic.

9.7. Avenues for reflection and application in current clinical practice

Space research is not disconnected from the terrestrial world. Thus, numerous scientific campaigns (publications and conferences) present the theme of "Life in Space for Life on Earth". As we described above, a parallel exists between a stay in microgravity and a sedentary lifestyle, severe inactivity and therefore accelerated aging. In a modern society in 2024, where a sedentary lifestyle represents a public-health scourge, professional and leisure activities are primarily carried out in a seated position. Yet, "sitting kills!" As Joan Vernikos perfectly reminds us in "Sitting Kills, Moving Heals: How Everyday Movement Will Prevent Pain, Illness, and Early Death," a title that evokes the need for regularity and multiplicity of gravitational stimuli and mobility throughout life to prevent pain, illness and even early death (Vernikos 2011). The teachings of space medicine are clearly in favor of combating the scourge of the effects of a sedentary lifestyle.

Furthermore, good adaptation to our environment involves the ability to efficiently integrate gravitational information in order to build and maintain motor maps allowing us to make high-quality anticipatory postural adjustments. These qualitative APAs will help minimize our energy expenditure during our daily activities. Regularly engaging the various sensory inputs implemented in the verticality model is therefore partly a guarantee for our good terrestrial bipedal health.

Our proprioceptive skills are the most prone to deterioration during inactivity. Indeed, proprioception is a real "sense of movement" (Berthoz 1997), which is therefore driven only by dynamic activity mobilizing all of our bodily segments in an environment that is, moreover, rich in stimuli. A simple experiment to demonstrate this phenomenon consists of placing your hand at rest, out of sight, without any activity of the upper limb for two minutes. You will then have to actively mobilize your fingers to determine their position without seeing them after this length of time. In addition, a sedentary lifestyle will lead to an accelerated degradation of the various components of proprioception, with the emergence of amyotrophy, the loss of joint mobility and amplitude and a slowing in the updating of our cortical motor maps, enabling high-performance APAs to be maintained.

Our sedentary lifestyle is often coupled with extensive screen use, exacerbating ocular vergence disorders, which can lead to harmful repercussions on our vestibulospinal tone. This element can be responsible for cervicalgia or recurrent migraines (Pérez-Cabezas et al. 2020). Furthermore, screens, which are often small in size, provide very little stimulation to our peripheral retina, which is indispensable for good postural control. Finally, in a static condition, sitting or standing in front of a screen, our fixed head is not subject to any acceleration capable of sensitizing our vestibular semicircular canals. Sedentary lifestyle coupled with the overuse of screens would therefore exacerbate the weight of visual input in the mix of sensory afferents building our sense of verticality.

These elements are not only theoretical: regular discussions with primary-school teachers highlight a worrying decline in motor performance among increasingly sedentary young children. This feeling should be linked to Lisa E. Bolger's systematic review of the motor skills of children around the world (Bolger et al. 2021). This element is associated with an increase in the number of overweight children, reaching more than 20% in developed countries, according to a WHO study (Spinelli et al. 2021). It should be recalled that due to the gravitational constraint, being overweight is one of the primary factors in osteoarthritis deterioration of the joints of the lower limbs, especially the knees (Reyes et al. 2016).

9.8. Practical consequences

We can retain the project of developing sophisticated preventive medicine, raising awareness as to the need for multiple solicitations and sensory stimuli.

Several aspects seem important to develop in an integrative manner. To this end, let us consider several actions to be implemented in current practice in the healthcare professions:

– Therapeutic and preventive education: combating in particular the scourge of a sedentary lifestyle (Vernikos 2011) and more physical and sports activities at school.

– The active postural concept of regular self-practiced antigravity spinal exercising (Kinel et al. 2021).

– Developing person-centered medicine focusing on empowerment when it comes to an individual's health, by giving them the power to self-manage (De Boer and Aydin 2023; Paakkari et al. 2023).

– Prevention of the deleterious effects of disc herniation by preserving the active balance in flexion/extension in favor of the extensor muscles (antigravity by definition), progression in the bearing of loads, especially heavy, particularly after a period of deconditioning (e.g. due to prolonged bed rest following surgery). Greater prevention of musculoskeletal disorders in the workplace, starting in school, offers all the more health messages to be promoted for the benefit of all.

– Planning for increased monitoring of ocular vergence disorders by incorporating more regular orthoptic checks.

– Preventing falls; proprioceptive work by varying conditions and solicitations. Notably performing exercises by removing the visual input in order to maximize the proprioceptive work of postural control regulation.

– Developing augmented virtual reality in medicine (Yeung et al. 2021).

– Working on people's representations and beliefs regarding health and illness to build self-confidence and confidence in their own ability to manage their health and illness.

– Increasing the level of health literacy in the general population in order to reduce the risk of disease chronicization and allowing for better communication between health-care professionals (Paakkari et al. 2023).

– Lifelong health, by combating a sedentary lifestyle and by promoting an active *modus vivendi* rich in sensorial solicitations. Using gravity as "our best sports coach" (Kourtidou-Papadeli et al. 2008).

– Maintaining mobility, active and passive joint and myofascial amplitude, throughout the whole body.

– Regular, progressive mechanical solicitations of an intensity adapted to individual constraints.

– Taking into account the person's capacity for recovery and its variable character.

NOTE.– More generally, the goal would be to think about health by considering human physiology in a complex relationship with its sensory, mechanical, psychological and social environment (Figure 9.5).

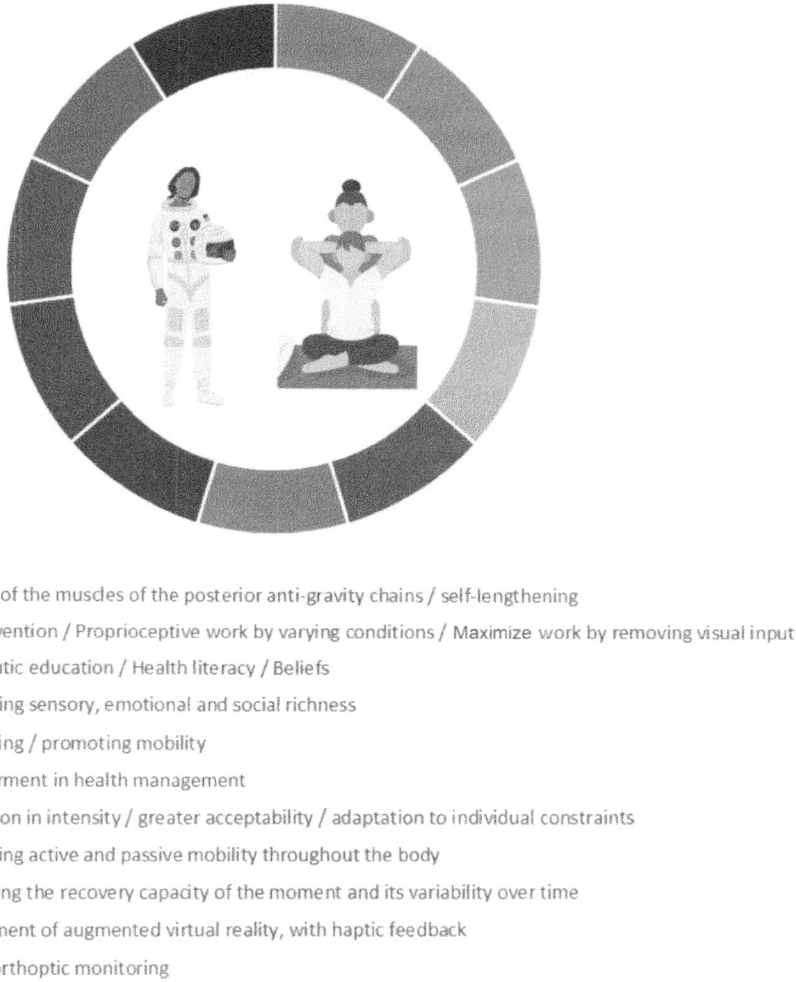

Figure 9.5. *Synthesis of actions in clinical practice of rehabilitation and manual osteopathic medicine, considering human physiology in a complex relationship with its sensory, mechanical, psychological and social environment. For a color version of this figure see www.iste.co.uk/dambricourtmalasse/futureofsapiens.zip*

9.9. Conclusion

Space medicine teaches us about aging and the consequences of a sedentary lifestyle or inactivity on Earth. Drawing on these lessons, it would be appropriate to build and strengthen integrative medicine, i.e. integrating allopathic and complementary medicine, to ultimately build medicine capable of raising awareness, preventing, accompanying and durably treating the deleterious effects of deconditioning in all its complexity. These challenges are multiple: political, economic and more generally, relating to public health. Furthermore, space exploration would involve, in the near future, increasingly long flights with prolonged stays in microgravity. The morphology of *Homo sapiens* would come to be completely modified by this antigravity environment. The study of human physiology in the spatial field shows to what extent our body is shaped by terrestrial gravity. The health of *Homo sapiens* on Earth is, in essence, a health of verticality.

9.10. Acknowledgments

We wish to dedicate this chapter to the memory of André Maurice Ouaknine (1945–2023) for all that he contributed to the understanding of upright humans. We would also like to thank Professor Claude Gharib for all his contributions to space medicine and, more generally, the life sciences scientists at the Centre national d'études spatiales[3].

9.11. References

Afshinnekoo, E., Scott, R.T., MacKay, M.J., Pariset, E., Cekanaviciute, E., Barker, R., Gilroy, S., Hassane, D., Smith, S.M., Zwart, S.R. et al. (2021). Fundamental biological features of spaceflight: Advancing the field to enable deep-space exploration. *Cell*, 184(24), 6002. doi: 10.1016/j.cell.2021.11.008.

Barra, J., Marquer, A., Joassin, R., Reymond, C., Metge, L., Chauvineau, V., Pérennou, D. (2010). Humans use internal models to construct and update a sense of verticality. *Brain: A Journal of Neurology*, 133(12), 3552–3563. doi: 10.1093/brain/awq311.

Bassani, T., Casaroli, G., Galbusera, F. (2019). Dependence of lumbar loads on spinopelvic sagittal alignment: An evaluation based on musculoskeletal modeling. *PloS One*, 14(3), e0207997. doi: 10.1371/journal.pone.0207997.

Belavy, D.L., Adams, M., Brisby, H., Cagnie, B., Danneels, L., Fairbank, J., Hargens, A.R., Judex, S., Scheuring, R.A., Sovelius, R. et al. (2016). Disc herniations in astronauts: What causes them, and what does it tell us about herniation on earth? *European Spine Journal*, 25(1), 144–154. doi: 10.1007/s00586-015-3917-y.

Berthoz, A. (1997). *Le sens du mouvement*. Odile Jacob, Paris [Online]. Available at: https://www.cairn-sciences.info/le-sens-du-mouvement--9782738104571.htm.

Berthoz, A., Lacour, M., Soechting, J.F., Vidal, P.P. (1979). The role of vision in the control of posture during linear motion. *Progress in Brain Research*, 50, 197–209. doi: 10.1016/S0079-6123(08)60820-1.

3 French national center for space studies.

Bertrand-Charette, M., Perron, M.-P., da Silva, R.A., Beaulieu, L.-D. (2023). Vibration-induced postural reactions: A scoping review on parameters and populations studied. *Frontiers in Human Neuroscience*, 17, 1307639. doi: 10.3389/fnhum.2023. 1307639.

Bolger, L.E., Bolger, L.A., O'Neill, C., Coughlan, E., O'Brien, W., Lacey, S., Burns, C., Bardid, F. (2021). Global levels of fundamental motor skills in children: A systematic review. *Journal of Sports Sciences*, 39(7), 717–753. doi: 10.1080/02640414.2020.1841405.

Bouisset, S. and Zattara, M. (1987). Biomechanical study of the programming of anticipatory postural adjustments associated with voluntary movement. *Journal of Biomechanics*, 20(8), 735–742. doi: 10.1016/0021-9290(87)90052-2.

Bubka, A. and Bonato, F. (2010). Natural visual-field features enhance vection. *Perception*, 39(5), 627–635. doi: 10.1068/p6315.

Buckey, J.C. (2006). *Space Physiology*. Oxford University Press, Oxford.

Buckley, J.G., Anand, V., Scally, A., Elliott, D.B. (2005). Does head extension and flexion increase postural instability in elderly subjects when visual information is kept constant? *Gait Posture*, 21(1), 59–64. doi: 10.1016/j.gaitpost.2003.11.005.

Carriot, J., Mackrous, I., Cullen, K.E. (2021). Challenges to the vestibular system in space: How the brain responds and adapts to microgravity. *Frontiers in Neural Circuits*, 15, 760313. doi: 10.3389/fncir.2021.760313.

Carvil, P.A., Attias, J., Evetts, S.N., Waldie, J.M., Green, D.A. (2017). The effect of the gravity loading countermeasure skinsuit upon movement and strength. *Journal of Strength and Conditioning Research*, 31(1), 154–161. doi: 10.1519/JSC.0000000000001460.

Chang, D.G., Healey, R.M., Snyder, A.J., Sayson, J.V., Macias, B.R., Coughlin, D.G., Bailey, J.F., Parazynski, S.E., Lotz, J.C., Hargens, A.R. (2016). Lumbar spine paraspinal muscle and intervertebral disc height changes in astronauts after long-duration spaceflight on the International Space Station. *SPINE*, 1. doi: 10.1097/BRS.0000000000001873.

Chen, W., Chao, J.-G., Zhang, Y., Wang, J.-K., Chen, X.-W., Tan, C. (2017). Orientation preferences and motion sickness induced in a virtual reality environment. *Aerospace Medicine and Human Performance*, 88(10), 903–910. doi: 10.3357/AMHP.4817.2017.

Clément, G. (2003). A review of the effects of space flight on the asymmetry of vertical optokinetic and vestibulo-ocular reflexes. *Journal of Vestibular Research: Equilibrium Orientation*, 13(4/6), 255–263.

Clément, G. and Wood, S.J. (2014). Rocking or rolling – Perception of ambiguous motion after returning from space. *PloS One*, 9(10). doi: 10.1371/journal.pone.0111107.

Cuisinier, R., Olivier, I., Vaugoyeau, M., Nougier, V., Assaiante, C. (2011). Reweighting of sensory inputs to control quiet standing in children from 7 to 11 and in adults. *PloS One*, 6(5), e19697. doi: 10.1371/journal.pone.0019697.

Custaud, M.A., Blanc, S., Gauquelin-Koch, G., Gharib, C. (2020). L'humain et l'espace : ses adaptations physiologiques [Online]. Available at: https://www.librairiemartelle.com/livre/9782322270620-l-humain-et-l-espace-ses-adaptations-physiologiques-marc-antoine-custaud-stephane-blanc-guillemette-gauquelin-koch-claude-gharib/.

Dakin, C. J. and Rosenberg, A. (2018). Gravity estimation and verticality perception. In *Handbook of Clinical Neurology*, Day, B.L. and Lord, S.R. (eds). Elsevier, Amsterdam, 159, 43–59.

De Abreu, S., Amirova, L., Murphy, R., Wallace, R., Twomey, L., Gauquelin-Koch, G., Raverot, V., Larcher, F., Custaud, M.-A., Navasiolava, N. (2017). Multi-system deconditioning in 3-day dry immersion without daily raise. *Frontiers in Physiology*, 8. doi: 10.3389/fphys.2017.00799.

De Boer, B. and Aydin, C. (2023). Empowerment: Freud, Canguilhem and Lacan on the ideal of health promotion. *Medicine, Health Care and Philosophy*, 26(3), 301–311. doi: 10.1007/s11019-023-10145-z.

Demangel, R., Treffel, L., Py, G., Brioche, T., Pagano, A.F., Bareille, M.-P., Beck, A., Pessemesse, L., Candau, R., Gharib, C. et al. (2017). Early structural and functional signature of 3-day human skeletal muscle disuse using the dry immersion model. *The Journal of Physiology*, 595(13), 4301–4315. doi: 10.1113/JP273895.

Dichgans, J. and Brandt, T. (1978). Visual-vestibular interaction: Effects on self-motion perception and postural control. In *Perception*, Held, R., Liebowitz, H.W., Teuber, H.L. (eds). Springer, Berlin, Heidelberg, 755–804. doi: 10.1007/978-3-642-46354-9_25.

Fitzpatrick, R. and McCloskey, D.I. (1994). Proprioceptive, visual and vestibular thresholds for the perception of sway during standing in humans. *The Journal of Physiology*, 478(1), 173–186. doi: 10.1113/jphysiol.1994.sp020240.

Fitzpatrick, R. and Watson, S.R.D. (2015). Passive motion reduces vestibular balance and perceptual responses. *The Journal of Physiology*, 593(10), 2389–2398. doi: 10.1113/JP270334.

Garrett-Bakelman, F.E., Darshi, M., Green, S.J., Gur, R.C., Lin, L., Macias, B.R., McKenna, M.J., Meydan, C., Mishra, T., Nasrini, J. et al. (2019). The NASA Twins Study: A multidimensional analysis of a year-long human spaceflight. *Science*, 364(6436), eaau8650. doi: 10.1126/science.aau8650.

Gibson, J.J. (2014). *The Ecological Approach to Visual Perception*. Psychology Press, New York. doi: 10.4324/9781315740218.

Gibson, J.J., Olum, P., Rosenblatt, F. (1955). Parallax and perspective during aircraft landings. *The American Journal of Psychology*, 68(3), 372–385.

Green, D.A. and Scott, J.P.R. (2017). Spinal health during unloading and reloading associated with spaceflight. *Frontiers in Physiology*, 8, 1126. doi: 10.3389/fphys.2017.01126.

Greene, K.A., Tooze, J.A., Lenchik, L., Weaver, A.A. (2022). Change in lumbar muscle size and composition on MRI with long-duration space flight. *Annals of Biomedical Engineering*, 50(7), 816–824. doi: 10.1007/s10439-022-02968-3.

Gunga, H.-C. (2015). *Human Physiology in Extreme Environments*. Academic Press, New York.

Hidalgo, D.F., MacMillan, A., Thomson, O.P. (2024). It's all connected, so it all matters – The fallacy of osteopathic anatomical possibilism. *International Journal of Osteopathic Medicine*, 100718. doi: 10.1016/j.ijosm.2024.100718.

Hides, J., Lambrecht, G., Ramdharry, G., Cusack, R., Bloomberg, J., Stokes, M. (2017). Parallels between astronauts and terrestrial patients – Taking physiotherapy rehabilitation "To infinity and beyond". *Musculoskeletal Science and Practice*, 27, S32–S37. doi: 10.1016/j.msksp.2016.12.008.

Hoch, M.C. and Russell, D.M. (2016). Plantar cooling does not affect standing balance: A systematic review and meta-analysis. *Gait Posture*, 43, 1–8. doi: 10.1016/j.gaitpost. 2015.10.011.

Hughson, R.L., Robertson, A.D., Arbeille, P., Shoemaker, J.K., Rush, J.W.E., Fraser, K.S., Greaves, D.K. (2016). Increased postflight carotid artery stiffness and inflight insulin resistance resulting from 6-mo spaceflight in male and female astronauts. *American Journal of Physiology. Heart and Circulatory Physiology*, 310(5), H628–H638. doi: 10.1152/ajpheart.00802.2015.

Keller, N., Whittle, R.S., McHenry, N., Johnston, A., Duncan, C., Ploutz-Snyder, L., Torre, G.G.D.L., Sheffield-Moore, M., Chamitoff, G., Diaz-Artiles, A. (2022). Virtual reality "exergames": A promising countermeasure to improve motivation and restorative effects during long duration spaceflight missions. *Frontiers in Physiology*, 13, 932425. doi: 10.3389/fphys.2022.932425.

Kenny, H.C., Rudwill, F., Breen, L., Salanova, M., Blottner, D., Heise, T., Heer, M., Blanc, S., O'Gorman, D.J. (2017). Bed rest and resistive vibration exercise unveil novel links between skeletal muscle mitochondrial function and insulin resistance. *Diabetologia*. doi: 10.1007/s00125-017-4298-z.

Kerstman, E. and Scheuring, R. (2011). NASA musculoskeletal space medicine and reconditioning program. In *83rd AsMA Annual Scientific Meeting*, Atlanta [Online]. Available at: https://ntrs.nasa.gov/citations/20110020319.

Kerstman, E., Scheuring, R.A., Barnes, M.G., DeKorse, T.B., Saile, L.G. (2012). Space adaptation back pain: A retrospective study. *Aviation, Space, and Environmental Medicine*, 83(1), 2–7.

Kim, J., Palmisano, S., Bonato, F. (2012). Simulated angular head oscillation enhances vection in depth. *Perception*, 41(4), 402–414. doi: 10.1068/p6919.

Kinel, E., D'Amico, M., Roncoletta, P. (2021). 3D quantitative evaluation of posture and spine proprioceptive perception through instinctive self-correction maneuver in adolescent idiopathic scoliosis. *Frontiers in Bioengineering and Biotechnology*, 9, 663394. doi: 10.3389/fbioe.2021.663394.

Kourtidou-Papadeli, C., Papadelis, C.L., Vernikos, J., Bamidis, P.D., Hitoglou-Antoniadou, M., Perantoni, E., Vlachogiannis, E. (2008). The therapeutic benefits of gravity in space and on earth. *Hippokratia*, 12(1), 28–31.

Lackner, J.R. and Dizio, P. (2005). Vestibular, proprioceptive, and haptic contributions to spatial orientation. *Annual Review of Psychology*, 56, 115–147. doi: 10.1146/annurev.psych.55.090902.142023.

Lê, T.-T. and Kapoula, Z. (2006). Distance impairs postural stability only under binocular viewing. *Vision Research*, 46(21), 3586–3593. doi: 10.1016/j.visres.2006.06.018.

Leach, M.J., Sundberg, T., Fryer, G., Austin, P., Thomson, O.P., Adams, J. (2019). An investigation of Australian osteopaths' attitudes, skills and utilisation of evidence-based practice: A national cross-sectional survey. *BMC Health Services Research*, 19(1), 498. doi: 10.1186/s12913-019-4329-1.

Liphardt, A.-M., Fernandez-Gonzalo, R., Albracht, K., Rittweger, J., Vico, L. (2023). Musculoskeletal research in human space flight – Unmet needs for the success of crewed deep space exploration. *NPJ Microgravity*, 9, 9. doi: 10.1038/s41526-023-00258-3.

Loehr, J.A., Lee, S.M.C., English, K.L., Sibonga, J., Smith, S.M., Spiering, B.A., Hagan, R.D. (2011). Musculoskeletal adaptations to training with the advanced resistive exercise device. *Medicine and Science in Sports and Exercise*, 43(1), 146–156. doi: 10.1249/MSS.0b013e3181e4f161.

Loehr, J.A., Guilliams, M.E., Petersen, N., Hirsch, N., Kawashima, S., Ohshima, H. (2015). Physical training for long-duration spaceflight. *Aerospace Medicine and Human Performance*, 86(12), A14–A23. doi: 10.3357/AMHP.EC03.2015.

Macias, B.R., Ferguson, C.R., Patel, N., Gibson, C., Samuels, B.C., Laurie, S.S., Lee, S.M.C., Ploutz-Snyder, R., Kramer, L., Mader, T.H. et al. (2021). Changes in the optic nerve head and choroid over 1 year of spaceflight. *JAMA Ophthalmology*, 139(6), 663–667. doi: 10.1001/jamaophthalmol.2021.0931.

Martin Paez, Y., Mudie, L.I., Subramanian, P.S. (2020). Spaceflight associated neuro-ocular syndrome (sans): A systematic review and future directions. *Eye and Brain*, 12, 105–117. doi: 10.2147/EB.S234076.

McDonnell, M.N. and Hillier, S.L. (2015). Vestibular rehabilitation for unilateral peripheral vestibular dysfunction. *The Cochrane Database of Systematic Reviews*, 1, CD005397. doi: 10.1002/14651858.CD005397.pub4.

McNamara, K.P., Greene, K.A., Moore, A.M., Lenchik, L., Weaver, A.A. (2019). Lumbopelvic muscle changes following long-duration spaceflight. *Frontiers in Physiology*, 10, 627. doi: 10.3389/fphys.2019.00627.

Mohammadi, M., Ghamkhar, L., Alizadeh, A., Shaabani, M., Salavati, M., Kahlaee, A.H. (2021). Comparison of the reliance of the postural control system on the visual, vestibular and proprioceptive inputs in chronic low back pain patients and asymptomatic participants. *Gait Posture*, 85, 266–272. doi: 10.1016/j.gaitpost.2021.02.010.

Moore, S.T., Clément, G., Dai, M., Raphan, T., Solomon, D., Cohen, B. (2003). Ocular and perceptual responses to linear acceleration in microgravity: Alterations in otolith function on the COSMOS and Neurolab flights. *Journal of Vestibular Research: Equilibrium Orientation*, 13(4/6), 377–393.

Norsk, P., Barr, Y., Clément, G., Cromwell, R., Fomina, E., Hirsch, N., Johnson-Green, P., Petersen, N., Scott, J., Tomilovskaya, E. et al. (2021). *Effectiveness of Physiological Countermeasures for Spaceflight Induced Deconditioning*. International Academy of Astronautics, Paris.

Ogle, K.N. (1951). The perception of the visual world. In *Science*, Gibson, J.J. and Carmichael, L. (eds). Houghton Mifflin, Boston, 113(2940), 535. doi: 10.1126/science.113.2940.535.a.

Oh, D.S. and Choi, J.D. (2021). Effects of motor imagery training on balance and gait in older adults: A randomized controlled pilot study. *International Journal of Environmental Research and Public Health*, 18(2), 650. doi: 10.3390/ijerph18020650.

Paakkari, L., Balch-Crystal, E., Manu, M., Ruotsalainen, J., Salminen, J., Ulvinen, E., Torppa, M. (2023). Health-literacy education drives empowerment and agency. *Lancet*, 401(10374), 343–344. doi: 10.1016/S0140-6736(23)00107-1.

Palmisano, S., Gillam, B.J., Blackburn, S.G. (2000). Global-perspective jitter improves vection in central vision. *Perception*, 29(1), 57–67. doi: 10.1068/p2990.

Paloski, W.H., Wood, S.J., Feiveson, A.H., Black, F.O., Hwang, E.Y., Reschke, M.F. (2006). Destabilization of human balance control by static and dynamic head tilts. *Gait Posture*, 23(3), 315–323. doi: 10.1016/j.gaitpost.2005.04.009.

Pérez-Cabezas, V., Ruiz-Molinero, C., Jimenez-Rejano, J.J., Chamorro-Moriana, G., Gonzalez-Medina, G., Chillon-Martinez, R. (2020). Effectiveness of an eye-cervical re-education program in chronic neck pain: A randomized clinical trial. *Evidence-Based Complementary and Alternative Medicine: eCAM*, 2760413. doi: 10.1155/2020/2760413.

Petersen, N., Lambrecht, G., Scott, J., Hirsch, N., Stokes, M., Mester, J. (2017). Postflight reconditioning for European astronauts – A case report of recovery after six months in space. *Musculoskeletal Science and Practice*, 27, S23–S31. doi: 10.1016/j.msksp.2016.12.010.

Reyes, C., Leyland, K.M., Peat, G., Cooper, C., Arden, N.K., Prieto-Alhambra, D. (2016). Association between overweight and obesity and risk of clinically diagnosed knee, hip, and hand osteoarthritis: A population-based cohort study. *Arthritis Rheumatology*, 68(8), 1869–1875. doi: 10.1002/art.39707.

Robin, A., Navasiolava, N., Gauquelin-Koch, G., Gharib, C., Custaud, M.-A., Treffel, L. (2022). Spinal changes after 5-day dry immersion as shown by magnetic resonance imaging (DI-5-CUFFS*). American Journal of Physiology. Regulatory, Integrative and Comparative Physiology*, 323(3), R310–R318. doi: 10.1152/ajpregu.00055.2022.

Roll, J.P., Vedel, J.P., Ribot, E. (1989). Alteration of proprioceptive messages induced by tendon vibration in man: A microneurographic study. *Experimental Brain Research*, 76(1), 213–222. doi: 10.1007/BF00253639.

Rossato, M., Ouaknine, M., Bourgeois, P. (2013). *Stabilometry Standard Guidelines 2011–2013 During Clinical Practice*. Marrapese Edizioni, Rome.

Saruco, E., Rienzo, F.D., Nunez-Nagy, S., Rubio-Gonzalez, M.A., Jackson, P.L., Collet, C., Saimpont, A., Guillot, A. (2017). Anodal tDCS over the primary motor cortex improves motor imagery benefits on postural control: A pilot study. *Scientific Reports*, 7(1), 480. doi: 10.1038/s41598-017-00509-w.

Sauer, A.K., Vigouroux, M., Dougherty, P.M., Cata, J.P., Ingelmo, P.M. (2023). Pain experience and sensory changes in astronauts during and after short-lasting commercial spaceflight: A proof-of-concept study. *Journal of Pain Research*, 16, 4253–4266. doi: 10.2147/JPR.S440630.

Severinsen, M.C.K. and Pedersen, B.K. (2020). Muscle-organ crosstalk: The emerging roles of myokines. *Endocrine Reviews*, 41(4), 594–609. doi: 10.1210/endrev/bnaa016.

Shen, M. and Frishman, W.H. (2019). Effects of spaceflight on cardiovascular physiology and health. *Cardiology in Review*, 27(3), 122. doi: 10.1097/CRD.0000000000000236.

Sherrington, C.S. (1907). Strychnine and reflex inhibition of skeletal muscle. *The Journal of Physiology*, 36(2/3), 185–204. doi: 10.1113/jphysiol.1907.sp001228.

Spinelli, A., Buoncristiano, M., Nardone, P., Starc, G., Hejgaard, T., Júlíusson, P.B., Fismen, A.-S., Weghuber, D., MusićMilanović, S., García-Solano, M. et al. (2021). Thinness, overweight, and obesity in 6- to 9-year-old children from 36 countries: The World Health Organization European Childhood Obesity Surveillance Initiative-COSI 2015–2017. *Obesity Reviews: An Official Journal of the International Association for the Study of Obesity*, 22(6), e13214. doi: 10.1111/obr.13214.

Stokes, M., Evetts, S., Hides, J. (2017). Terrestrial neuro-musculoskeletal rehabilitation and astronaut reconditioning: Reciprocal knowledge transfer. *Musculoskeletal Science and Practice*, 27, S1–S4. doi: 10.1016/j.math.2016.10.003.

Treffel, L. (2017). Dysfonctions vertébrales et posturales après simulations de la microgravité. PhD Thesis, Université de Strasbourg, Strasbourg [Online]. Available at: https://theses.hal.science/tel-01702866.

Treffel, L., Dmitrieva, L., Gauquelin-Koch, G., Custaud, M.-A., Blanc, S., Gharib, C., Millet, C. (2016). Craniomandibular system and postural balance after 3-day dry immersion. *PloS One*, 11(2), e0150052. doi: 10.1371/journal.pone.0150052.

Tulimieri, D.T. and Semrau, J.A. (2023). Aging increases proprioceptive error for a broad range of movement speed and distance estimates in the upper limb. *Frontiers in Human Neuroscience*, 17, 1217105. doi: 10.3389/fnhum.2023.1217105.

Vernikos, J. (2011). *Sitting Kills, Moving Heals: How Everyday Movement Will Prevent Pain, Illness, and Early Death – And Exercise Alone Won't*. Quill Driver Books, Fresno, CA.

Vernikos, J. and Schneider, V.S. (2010). Space, gravity and the physiology of aging: Parallel or convergent disciplines? A mini-review. *Gerontology*, 56(2), 157–166. doi: 10.1159/000252852.

Von Schoen-Angerer, T., Manchanda, R.K., Lloyd, I., Wardle, J., Szöke, J., Benevides, I., Martinez, N.S.A., Tolo, F., Nicolai, T., Skaling-Klopstock, C. et al. (2023). Traditional, complementary and integrative healthcare: Global stakeholder perspective on WHO's current and future strategy. *BMJ Global Health*, 8(12), e013150. doi: 10.1136/bmjgh-2023-013150.

Wasicky, R., Ziya-Ghazvini, F., Blumer, R., Lukas, J.R., Mayr, R. (2000). Muscle fiber types of human extraocular muscles: A histochemical and immunohistochemical study. *Investigative Ophthalmology Visual Science*, 41(5), 980–990.

de Winkel, K.N., Katliar, M., Diers, D., Bülthoff, H.H. (2018). Causal inference in the perception of verticality. *Scientific Reports*, 8(1), 5483. doi: 10.1038/s41598-018-23838-w.

de Winkel, K.N., Talsma, T.M.W., Happee, R. (2022). A meta-analysis of simulator sickness as a function of simulator fidelity. *Experimental Brain Research*, 240(12), 3089–3105. doi: 10.1007/s00221-022-06485-6.

Winnard, A., Nasser, M., Debuse, D., Stokes, M., Evetts, S., Wilkinson, M., Hides, J., Caplan, N. (2017). Systematic review of countermeasures to minimise physiological changes and risk of injury to the lumbopelvic area following long-term microgravity. *Musculoskeletal Science and Practice*, 27, S5–S14. doi: 10.1016/j.msksp.2016.12.009.

Yelnik, A. (2021). *Troubles de l'équilibre : aspects sensoriels de la physiologie à la rééducation*. Elsevier, Issy-les-Moulineaux.

Yeung, A.W.K., Tosevska, A., Klager, E., Eibensteiner, F., Laxar, D., Stoyanov, J., Glisic, M., Zeiner, S., Kulnik, S.T., Crutzen, R. et al. (2021). Virtual and augmented reality applications in medicine: Analysis of the scientific literature. *Journal of Medical Internet Research*, 23(2), e25499. doi: 10.2196/25499.

Yuan, M., Custaud, M.-A., Xu, Z., Wang, J., Yuan, M., Tafforin, C., Treffel, L., Arbeille, P., Nicolas, M., Gharib, C. et al. (2019). Multi-system adaptation to confinement during the 180-day controlled ecological life support system (CELSS) experiment. *Frontiers in Physiology*, 10, 575. doi: 10.3389/fphys.2019.00575.

Zhang, L.-F. and Hargens, A.R. (2018). Spaceflight-induced intracranial hypertension and visual impairment: Pathophysiology and countermeasures. *Physiological Reviews*, 98(1), 59–87. doi: 10.1152/physrev.00017.2016.

Conclusion

The Future of *Sapiens* and Verticality: Cerebro-Cerebellization, AI and Emotion

The nine chapters of this book open up perspectives for shared reflection on the future of our biological species, confronted with a phylogenetic destiny, that of an internal physical limit with the verticality of the central nervous system and the unstable balance of the body in a vertical position. Research on living conditions in extraterrestrial space shows that, while the hominization of the body occurs in an aqueous medium, therefore close to weightlessness (physico-chemical phylogenetic determinisms); while it is first embryonic; and finally, while humanization is already intrauterine (relational sensitivity), the postnatal pursuit of this hominization/ humanization process is evidently conditioned by Earth's gravitational forces. *Sapiens* is, biologically, a terrestrial phenomenon that appears ill-adapted to extraterrestrial physical conditions in weightlessness.

It still remains to be known how to interpret the meaning to be given to a natural phylogenetic destiny, according to the importance that contemporary societies intend to give to the emerging effects of the growing complexity of the central nervous system. The most tangible, in more than 3 million years, are the lithic productions of an ever more inventive technical intelligence, liberating from environmental constraints and succeeding in overcoming the limits of primates' familiar environments: maritime, geomorphological and climatic, for seabeds, high altitude, the conquest of the poles and space exploration. These limits have become ever more extreme, oriented towards those of the infinitely small (quantum mechanics), those of the infinitely large (age and size of the universe in expansion) and those of a third virtual axis that came to be established as an observation, but not before the

Conclusion written by Anne DAMBRICOURT MALASSÉ, Sandra JOFFROY, Djillali HADJOUIS and Bernard AUTET.

middle of the 20th century, with the paleontologist and geologist Pierre Teilhard de Chardin (1881–1955), the arrow of space-time.

This axis is virtual, as it unfolds over the long, now-disappeared cosmogenic periods. It is reflected in the growing complexification of atomic and then molecular assemblies, endowed with properties of self-organization, self-reproduction and therefore memorization. Planet Earth, virtually oceanic 3.2 billion years ago, allowed these properties to organize themselves into dynamic systems closed in on themselves but open to their environment and to be transmitted. These were the first single-cell cycles or the emergence of life. Paleontologist academicians, prehistorians and ethnologists recognize that this third virtual axis has never ceased to transmit its properties of internal complexifications and that the most complex organization thus generated by this terrestrial phenomenon is the human brain. The brain is recognized, without question, as the most complex system that terrestrial evolution has ever produced (Changeux 2014). Henceforth, it is the complexity of the verticalized central nervous system that will be focused upon, with the integration of the cerebellum connected with the cerebral neocortex. The nervous system of *Sapiens* is therefore the most complex of all biological systems, past and present. It is also the most self-reflective information reflection system that results in the emergence of consciousness. This self-reflective reflection has achieved, with *Sapiens*, anticipatory capabilities allowing, through abstract projection, consequences into the long term. Having become collective, it has already projected itself beyond the terrestrial sphere. This is why this virtual axis is considered to be that which tends towards the infinitely reflective or consciously reflected, until it covers the planetary surface, forming a new terrestrial envelope called the "noosphere". This phylogenesis, internal, inherited and geometrically oriented with chordal verticalization, is well-recognized. It is objective and awakens the sense of evolutionary collective responsibilities. Such are the observers of the late 20th century and/or the first quarter of the 21st century, certainly few in number as in any pioneering science, but essential figures, such as paleontologists-paleoanthropologists Jean Piveteau (1983) and Yves Coppens (2005), and prehistorians André Leroi-Gourhan (Schlanger 2023) and Henry de Lumley (1981, 2018):

> Without any break in continuity with what precedes it, the advent of mankind marks an entirely original stage, of equal importance to what was the appearance of life itself, and which can be defined as the establishment on the planet of a thinking sphere, superimposed on the biosphere, the noosphere. Within it, the immense effort of cerebralization that began on the juvenile Earth will be completed, moving toward collective organization or socialization (Piveteau 1965).

This technical manual intelligence was accompanied by an emotional intelligence and a symbolic imagination, as evidenced by the Makapansgat manuport pebble with a hominin-like face. This pebble was collected 3 million years ago in South Africa by a hominid drawn to the "reflection" of a face set in a natural pebble habitually gathered to make into a carved tool. The choice that guided the gesture is interpreted as the beginnings of an aesthetic sense, of artistic creation (Oakley 1981). However, 3 million years ago, it was a scavenger collecting pebbles to cut up the corpses that constituted its protein and fat resource while avoiding itself becoming a predator's prey. This considered and conscious gesture would instead reveal an appeal that arose during an activity focused on a practical necessity. The natural sculpture likely evoked an awareness of one's own image, breaking with routine behavior. The encounter between the hominid's gaze and this pebble was sufficiently troubling for the individual to decide to preserve it as it was and take it back to its place of residence. The reflection of this image interacted with the memory of its face, or that of a fellow hominid, or even the image of the skull of another, deceased, fellow hominid. This is why symbolic and emotional thought is considered the language of meaning, and art as its cultural, and even cultural, expression.

With this pebble, it is no longer a question of technical and practical intelligence but of a conscious reflection of the self that counted as much as, if not more than, the collecting of the stone to be carved. This choice reveals to the anthropologist the very great antiquity of an imaginative, sensitive and emotional reflexive consciousness, likely questioning and certainly introspective. Just as the carved pebble, or chopper, bears witness to the external life of the hominid, with its technical intelligence in relation to its territory, its dangers and its riches useful for everyday life, so the natural pebble with a hominid face teaches us about the intimate life, likely the questioning reflection on the identity of the self, otherness and the differences between the living in motion and the inert dead.

Anticipation, with the abstract projection of possibilities and the enactment of judicious choices, is one of the most salient emerging cognitive properties of the hominization process, since the cerebellum has learned as a priority to avoid its own falling by anticipating the movements to adopt. The will not to fall, like that of straightening up, is acquired with learning to walk.

At present, it is necessary to refocus on the paradigm of the internal complexification of the organization of life, with the many "left behind" in human paleontology, as this book shows: embryogenesis, the fetus, pregnant and parturient women, psychomotor development, occluso-postural balance, the cognitive properties of the neocerebellum and the construction of a symbolic image of self that manages, through otherness and recognition of the other, to overcome the injurious accidents of a natural evolution.

This refocusing on the course of growing complexity is set to become a new threshold in awareness of evolution as an authentic planetary fact that concerns us, because internal verticality is the unsurpassable physical limit, and it is from this limit, reached for the first time in the evolution of vertebrates, that the complexification will continue according to its own principles of self-organization. If it is not already the case with dental agenesis, the migration of the neural crest cells as well as the complexification of the connections of the central nervous system will change. The reduction of the dental formula from 32 to 20 teeth with one incisor, one canine, one premolar and two molars will necessarily change the occluso-postural balance; the neocerebellum and its neuronal connections will be solicited differently, and the image of the face will change; new disorders and new emotions will emerge, with symbolic expressions to express their meaning.

The role and place of artificial intelligence (AI) in the process of humanization in the making gave rise, between 1945 and 1970, to notable evolutionary reflections with Medawar, Julian Huxley, Hardin, Engglebart, Popper, Bateson and Licklider on the nature of thought, brain functions and links with cybernetics (Bates 2023). It is now necessary to think of them in the evolutionary context of a process of hominization inherited and in the making. At the moment of this realization, we are like the hominid of Makapansgat who stopped selecting pebbles while being guided by its routine behavior. The hominid likely experienced the shock of the image of a face staring back at it from the stone to be broken. The hominid did not smash it, because it was not a potential tool, instead taking it away and keeping it, probably struck by the disconcerting emotion it triggered.

Emotion has its place in this inherited process, and living in a certain way, since biological evolution has reproduced irreversibly with its self-organized mechanisms for 3.5 billion years, despite major extinction crises. The Anthropocene reminds our Western consciousness that our cultures are not the complex, misunderstood processes of evolution; AI even less so, deprived as a carved pebble is, of a consubstantial link with living evolution and its mechanisms of nervous-system complexification.

The growing interest in the crossroads of paleontology, biology and psychology, supported by the contributions of bioinformatics and AI, will undoubtedly lead to many further discoveries, primordial in understanding the central nervous system of humans and their relationship with their own verticality, which, at present, still constitutes an unknown. It is therefore favorable, and such is the ethical spirit that animates FREHOPS, to recall that AI, like carved pebbles, is integrated into the natural, social and cultural balances of humans with their ecosystems, and that it is contained in the biosphere, adding complexity within the parameters of adaptations and technical innovations of human societies. Its place in the humanization process is that of a tool, algorithmic assistance serving planetary, plant, animal and human health, exposed to the unpredictable effects of the complexification of anthropogenic, cultural and conflictual activities, combined with global warming.

Having reached the present time of the third virtual axis, it is a physical wall that emerges, the wall of the verticalization of the vertebrate organization that coincides with the Earth's gravitational field. Thus, contrary to the assertions of Jacques Monod (1970) in the midst of a cybernetic revolution: "The old alliance is broken. Man finally knows that he is alone in the indifferent immensity of the Universe from which he emerged by chance. Neither his destiny nor his duty are inscribed anywhere". Inherited destiny is a fact, since our phylogenetic identity is that of an irreversible amplification of effects of gametogenesis that are active from fertilization.

It is indeed a question of a destiny, and of inherited processes, with which conscious reflection must "form an alliance" to avoid the fall; in other words, the lack of well-being of those who have not yet been born. This is precisely what gives meaning to the concepts of collective evolutionary responsibility (Welinski 2021), as well as to anticipation capacities, with a probable cerebellization of encephalization, the development of social relations and emotional memory of which are the lasting effects of humanization.

References

Bates, D. (2023). L'amplification exosomatique de l'intelligence. Biologie, épistémologie, cybernétique. Partie I. Par-delà le cognitivisme : penser l'extériorisation de l'esprit. *Appareil*. doi: 10.4000/appareil.7138.

Changeux, J.-P. (2014). Le cerveau et la complexité. In *Le cerveau et la pensée, le nouvel âge des sciences cognitives*, Dortier, J.F. (ed.). Éditions Sciences Humaines, Paris, 93–101.

Coppens, Y. (2005). Préface. In *Teilhard de Chardin, visionnaire du monde nouveau*, Danzin, A. and Masurel, J. (eds). Editions du Rocher, Paris.

de Lumley, H. (1981). Pierre Teilhard de Chardin, paleontologist, philosopher, theologian. In *International Symposium on the Occasion of the Centenary of the Birth of Pierre Teilhard de Chardin*, 16–18 September, UNESCO, Paris.

de Lumley, H. (2018). De l'origine de l'Homme à l'ultra-humain de Pierre Teilhard de Chardin. In *Colloque de la fondation Institut de paléontologie humaine*. 30 November–1 December, Paris.

Monod, J. (1970). *Le hasard et la nécessité. Essai sur la philosophie naturelle de la biologie moderne*. Point, Paris.

Oakley, K.P. (1981). Emergence of higher thought. *Philosophical Transactions of the Royal Society of London B, Biological Sciences*, 292(1057), 205–211.

Piveteau, J. (1965). Préface. In *La place de l'homme dans la Nature. Le groupe zoologique humain*, Teilhard de Chardin, P. (ed.). Union Générale d'Editions, Paris.

Piveteau, J. (1983). *Origine et destinée de l'Homme*. Masson, Paris.

Schlanger, N. (2023). *L'invention de la technologie. Une histoire intellectuelle avec André Leroi–Gourhan*. PUF, Paris.

Welinski, M. (ed.) (2021). Entretiens avec Anne Dambricourt Malassé. In *Comment bien vivre la fin de ce monde. Les points de vue éclairants de quinze personnalités sur la question de l'effondrement de notre civilisation*. Guy Tredaniel Éditeur, Paris, 73–89.

List of Authors

Bernard AUTET
Fédération des Recherches sur l'Évolution Humaine, l'Ostéopathie et la Posture au service de la Santé
Pérols
France

July BOUHALLIER
Centre de Formation Professionnelle Continue en Ostéopathie
and
Institut de Recherche et d'Action pour la Santé des Femmes
Paris
France

Michel CLAUZADE
ODENTH
Sanary-sur-Mer
France

Numa CLAUZADE
ODENTH
Sanary-sur-Mer
France

Anne DAMBRICOURT MALASSÉ
UMR 7194 CNRS
Muséum national d'Histoire naturelle
Paris
France

Tien-Tuan DAO
UMR CNRS 9013
Laboratoire de Mécanique, Multiphysique, Multiéchelle
Centrale Lille
France

Catherine DOLTO
CIRDH Frans Veldman
Paris
France

Denis DUCOMMUN
Pôle Recherche
Institut Toulousain d'Ostéopathie
Labège
France

Djillali HADJOUIS
Union Internationale des Sciences Préhistoriques et Protohistoriques
Paris
France

Marie-Christine HO BA THO
Biomécanique et Bioingénierie
CNRS
Centre Royallieu
Université de technologie de Compiègne
France

Sandra JOFFROY
Institut Toulousain d'Ostéopathie
Labège
and
F2SMH
Université de Toulouse
France

Fabienne LALLOUET
Hôpital Américain de Paris
Neuilly-sur-Seine
France

Véronique LEROY-MALHERBE
SESSAD-APF France Handicap
Arpajon
France

Tan-Nhu NGUYEN
International University
Ho Chi Minh City
Vietnam

Élise PINEAU
SESSAD-APF France Handicap
Arpajon
France

Virginie PINEAU
SESSAD-APF France Handicap
Arpajon
France

Loïc TREFFEL
UTS-ARCCIM
University of Technology
Sydney
Australia

Index

E, F

G, H

L, M

N, O

P, Q

R, S

T, U

V, W, Z

Other titles from

iSTE

in

Biology and Biomedical Engineering

2024

MASQUIDA Benoît, LECLERC Fabrice
Looking at Ribozymes: Biology of Catalytic RNA
(Nucleic Acids Set – Volume 2)

NICOLAS Violaine
Inventory of Biodiversity Today: New Methods and Discoveries

WALCH Jean-Paul, BLAISE Solange
Phyllotaxis and Symmetry in Angiosperms

2023

BALTHAZART Jacques
Fifty Years of Evolution in Biological Research: Progress and Decline

GILBERT Marc, PIRKMAJER Sergej
Concepts in Biology: A Historical Perspective

VERNA Emeline
Asymptomatic Osseous Variations of the Postcranial Human Skeleton
(Comparative Anatomy and Posture of Animal and Human Set – Volume 5)

WALCH Jean-Paul, BLAISE Solange
Phyllotaxis Models: A Tool for Evolutionary Biologists

2022

DAMBRICOURT MALASSÉ Anne
Embryogeny and Phylogeny of the Human Posture 2: A New Glance at the Future of our Species
(Comparative Anatomy and Posture of Animal and Human Set – Volume 4)

FOSSÉ Philippe
Structures and Functions of Retroviral RNAs: The Multiple Facets of the Retroviral Genome
(Nucleic Acids Set – Volume 1)

2021

DAMBRICOURT MALASSÉ Anne
Embryogeny and Phylogeny of the Human Posture 1: A New Glance at the Future of our Species
(Comparative Anatomy and Posture of Animal and Human Set – Volume 3)

GRANDCOLAS Philippe, MAUREL Marie-Christine
Systematics and the Exploration of Life

HADJOUIS Djillali
The Skull of Quadruped and Bipedal Vertebrates: Variations, Abnormalities and Joint Pathologies
(Comparative Anatomy and Posture of Animal and Human Set – Volume 2)

HULLÉ Maurice, VERNON Philippe
The Terrestrial Macroinvertebrates of the Sub-Antarctic Îles Kerguelen and Île de la Possession

2020

CAZEAU Cyrille
Foot Surgery Viewed Through the Prism of Comparative Anatomy: From Normal to Useful
(Comparative Anatomy and Posture of Animal and Human Set – Volume 1)

DUJON Bernard, PELLETIER Georges
Trajectories of Genetics

2019

BRAND Gérard
Discovering Odors

BUIS Roger
Biology and Mathematics: History and Challenges

2016

CLERC Maureen, BOUGRAIN Laurent, LOTTE Fabien
Brain–Computer Interfaces 1: Foundations and Methods
Brain–Computer Interfaces 2: Technology and Applications

FURGER Christophe
Live Cell Assays: From Research to Health and Regulatory Applications

2015

CLARYSSE Patrick, FRIBOULET Denis
Multi-modality Cardiac Imaging: Processing and Analysis

2014

CHÈZE Laurence
Kinematic Analysis of Human Movement

DAO Tien Tuan, HO BA THO Marie-Christine
Biomechanics of the Musculoskeletal System: Modeling of Data Uncertainty and Knowledge

FANET Hervé
Medical Imaging Based on Magnetic Fields and Ultrasounds

FARIÑAS DEL CERRO Luis, INOUE Katsumi
Logical Modeling of Biological Systems

MIGONNEY Véronique
Biomaterials

TEBBANI Sihem, LOPES Filipa, FILALI Rayen, DUMUR Didier, PAREAU Dominique
CO_2 Biofixation by Microalgae: Modeling, Estimation and Control

Printed and bound by CPI Group (UK) Ltd, Croydon, CR0 4YY

28/07/2026

14931346-0003